Perspectives in
Behavioral Medicine

HEALTH AT WORK

Perspectives in Behavioral Medicine
Sponsored by the Academy of Behavioral Medicine Research

Perspectives in
Behavioral Medicine

HEALTH AT WORK

Edited by

Stephen M. Weiss
National Heart, Lung, & Blood Institute
National Institutes of Health

Jonathan E. Fielding
University of California, Los Angeles, and
Johnson & Johnson Health Management, Inc.

Andrew Baum
F. Edward Hébert School of Medicine
Uniformed Services University of the Health Sciences

LEA LAWRENCE ERLBAUM ASSOCIATES, PUBLISHERS
1991 Hillsdale, New Jersey   Hove and London

Lawrence Erlbaum Associates, Inc., Publishers
365 Broadway
Hillsdale, New Jersey 07642

Library of Congress Cataloging-in-Publication Data

Health at work / edited by Stephen M. Weiss, Jonathan E. Fielding,
   Andrew Baum.
       p.   cm. — (Perspectives in behavioral medicine)
     Includes bibliographical references.
     Includes indexes.
     ISBN 0–8058–0770–5
   1. Health promotion.   2. Industrial hygiene.   I. Weiss, Stephen
   M.   II. Fielding. Jonathan E.   III. Baum, Andrew.   IV. Series.
     [DNLM: 1. Health Promotion.   2. Occupational Health Services.   WA
   400 H43452]
   RC969.H43H433   1990
   658.3′82—dc20
   DNLM/DLC
   for Library of Congress   90–13810
                                                                   CIP

Printed in the United States of America
10 9 8 7 6 5 4 3 2 1

CONTENTS

LIST OF CONTRIBUTORS

DAVID R. ANDERSON   StayWell Health Management Systems, Inc., 1285 Corporate Center Drive, Suite 100, Eagan, MN 55121

MARSHALL H. BECKER   School of Public Health, University of Michigan, 109 South Observatory Street, Ann Arbor, MI 48109-2029

JOHANNES BRENGELMANN   Max-Planck Institute for Psychiatry, Department of Psychology, Kraepelinstr. 2, D-8000 Munchen 40, WEST GERMANY

KELLY D. BROWNELL   Department of Psychiatry, University of Pennsylvania, 133 South 36th Street, Philadelphia, PA 19104

RICHARD A. CARLETON   Memorial Hospital of Rhode Island, Brown University, Pawtucket, RI 02860

JONATHAN E. FIELDING   Johnson & Johnson Health Management, Inc., 2825 Santa Monica Blvd., Suite 200, Santa Monica, CA 90404

JOHN P. FOREYT   Department of Medicine, Baylor University College of Medicine, 6535 Fannin M.S. F700, Houston, TX 77030

NEIL E. GRUNBERG   Dept. of Medical Psychology, Uniformed Services University of the Health Sciences, 4301 Jones Bridge Road, Bethesda, MD 20814-4799

MARGARET HAMBURG   New York City Department of Health, 125 Worth Street, Room 338, New York, NY 10013

WILLIAM L. HASKELL   Stanford Center for Disease Prevention, 730 Welch Road, Suite B, Stanford, CA 94305

DOROTHEA JOHNSON   AT&T, 295 North Maple Avenue, Room 4A08E3, Basking Ridge, NJ 07920

WILLIAM S. JOSE II   OverView Consulting, 5500 West 66th Street, Edina, MN 55435

GEORGANNA LEAVESLEY   Department of Psychology, University of Houston

ROBERT W. LUTZ   School of Medicine, University of California

TERRY MASON   Health Improvement Programs, Johnson & Johnson Health Management, One Johnson & Johnson Plaza, New Brunswick, NJ 08933

KENNETH R. PELLETIER   School of Medicine, University of California

STEPHEN M. WEISS   National Heart, Lung, & Blood Institute, National Institutes of Health, Federal Bldg., Room 216, Bethesda, MD 20892

1 Health at Work

Stephen M. Weiss
National Heart, Lung, and Blood Institute
National Institutes of Health

The World Health Organization defines health as ". . . a state of complete physical, mental, and social well-being and not merely the absence of disease or infirmity" (World Health Organization, Preamble, 1948). This seemingly simple statement speaks to a paradigm shift in our thinking—a recognition that health cannot be defined by curative medicine alone, nor by discovery of causation of disease. Rather, health is considered a "state of being" which has social, economic, cultural, and political correlates as well as involving the physiological integrity of the organism.

Seeking solutions to health problems, therefore, becomes the province of multiple disciplines and interest groups working at multiple levels within "the system." The virtual conquest of the acute infectious disorders has left the chronic degenerative diseases (heart disease, cancer, and stroke) as the major causes of morbidity and mortality in the United States. Inasmuch as we all will eventually die from *something,* this fact in itself would not be disheartening were it not for the distressingly high levels of *premature* morbidity and mortality associated with these figures. Combined with accidents, the chronic diseases account for 75% of the premature mortality experienced in the United States (Beary, 1981).

Affecting the health status of a national population (not to mention a "world" population) is no small undertaking. Health has without question become a national priority in the United States, perhaps *the* national priority, as our concern about national defense becomes less pressing. Unfortunately, health expenditures appear determined to outstrip the defense establishment in rate of escalation, as well as size. From a national health expenditure of $10 billion in 1950, we are currently facing costs in excess of $660 billion today,

1

with a projected increase to $1.5 trillion by the year 2000 (HCFA, 1988). Even more important, the percentage of our Gross National Product (GNP) devoted to health is also increasing, from 4.4% in 1950 to an estimated 15% by 2000. Our continued reliance on "high tech" curative medicine will continue to escalate costs, perhaps at an even greater rate than predicted. Although we cannot ethically withhold lifesaving treatment for those in need, to ultimately exert effective cost control over health expenditures, we must "bite the bullet" and invest, at the same time, in *long*-term preventive strategies that will ultimately lead to a *lessening of need* for such curative services.

When we speak of prevention, we really mean *delay*. We have not learned to *prevent* death, but we are particularly concerned about the premature (i.e., under age 65) departures from this earth. Since 1900, we have seen an increase of 20 years in the lifespan of the average American. However, a closer inspection of these figures reveals the major changes have taken place during the childhood years, the result of immunization, penicillin, and other pharmacologic discoveries to combat the acute infectious diseases. For the 45-year-old male, the increase in lifespan since 1900 has amounted to only about 5 years. Thus, progress toward ameliorating the effects of the chronic degenerative diseases has left much to be desired. The goal, therefore, is not to prolong life past the stage where biological collapse would be expected (85 to 95 years of age) but to reduce morbidity prior to that time, with terminal events being preceded by relatively short illnesses (Fries & Crapo, 1981). Under those circumstances, one could expect enhanced quality of life, decreased reliance on the curative medical system and associated services, and containment of health costs to an affordable level for all concerned.

When we speak of prevention efforts, we use terms such as "risk factor" and "risk-factor reduction" to describe both the problem and its potential resolution. Epidemiological studies have identified "markers" and related behavior patterns which have demonstrated consistent associations with increased probability of disease (e.g., Belloc & Breslow, 1972; Keys, et al., 1971; Rosenman et al., 1975). The term *probability* is important to keep in mind as there is no guarantee that a given individual will remain healthy (or become ill) by following certain regimens or engaging in certain behaviors. Risk factors for disease are essentially "probability statements" that should be considered by the individual as information relevant to personal decision-making regarding health-enhancing (or inhibiting) behavior patterns.

From a public health and national health planning standpoint, however, developing risk-factor reduction strategies should be a *key* element in designing a comprehensive approach to improving the nation's health while containing and ultimately reducing national health expenditures. Taking a closer look at some of the more obvious risk factors, we find elevated blood pressure, smoking, high serum cholesterol, and the "coronary-prone" behavior pattern to be four of the most widely accepted risk factors for coronary heart disease and stroke.

Smoking, diet, and environmental pollutants are some of the risk factors associated with cancer, while substance abuse (particularly alcohol), seat belts, and utilization of available safety equipment on the job have obvious implications for accident prevention. The development of such programs must also focus on "health promotion" as a logical extension of the disease-prevention perspective. Although considerable overlap exists in defining the terms, health promotion also includes assisting well people to feel even better—perhaps better than they have ever felt in their lives. Therefore, one can speak of health enhancement without reference to disease (e.g., the WHO definition of health). Thus, in addition to the more quantifiable measures of utilization of the health care system, one should also address "quality of life" measures that have implications for a broad variety of issues associated with life and job satisfaction.

Health promotion programs have been developed within three major domains: (1) schools; (2) worksites; and (3) communities. Although all three have significant public health implications, the unique constellation of factors pertinent to the worksite merit special consideration.

First, the more than 110 million employed persons in our society are those at greatest risk for premature morbidity and mortality (almost all under age 65) and are the backbone of our society's productivity. The employers of these 110 million workers have a major stake in the overall welfare of their employees, as health status (and employee health costs) contribute in a major way to productivity and profitability (or lack thereof) of their enterprises. Absenteeism, for example, is a major source of lost productivity. Smokers are absent 5 days per year more than nonsmokers, accounting for 80 million workdays lost per year. The prospect of both reducing employee health costs *and* increasing the health status of the workforce is a powerful incentive to management.

As most employed persons have families, the inclusion (directly or indirectly) of those family members into the health promotion milieu increases both the coverage and potential benefit to perhaps another 100 million persons. Thus, in addition to the workforce, worksite programs have the potential to extend to a large proportion of the entire U.S. population.

Finally, the design, implementation, and evaluation of health promotion programs may involve considerable financial investment over several years before one can hope to see evidence of benefit (Doherty, 1988). The private sector has a long tradition of financing research and development to assure their long-term standing in the marketplace, so such concepts are not foreign to their traditional approaches to new ideas. They do need to have evidence, however, of the potential efficacy and benefit of these programs to justify commitment of corporate assets to such endeavors.

This book focuses on the major issues concerning the need for worksite health promotion programs, identifies and discusses examples of the most intensively studied programs (e.g., Johnson & Johnson's "Live for Life";

AT&T's "TLC"; Control Data's "Staywell"; and the Coors Wellness Program) and considers the "State of Science" for the four most frequently offered health-promotion program components:

1. Smoking cessation;
2. Weight control;
3. Exercise;
4. Stress management.

Finally, the major challenges facing worksite programs will be addressed:

- Program Development and Design;
- Cost Benefit/Cost Effectiveness;
- Legislative/Policy Issues;
- The Relationship of Community Health Promotion Efforts to Worksite Programs;
- Program Participation/Recruitment;
- Ethical Concerns.

In the first section, each author addresses salient aspects of the dilemmas and opportunities facing the field today, illustrating points from several of the largest and most carefully evaluated worksite health-promotion programs in the United States. Chapters by Jonathan Fielding and Dorothea Johnson address the key challenges facing both the purveyors of worksite health promotion and industry decision-makers. Dr. Fielding considers the prevalence, breadth (and quality) of health-promotion programs at the worksite, evaluation of efficiency and cost-benefit, and the key challenges that must be addressed by the field, using the Johnson and Johnson "Live for Life" program to illustrate his points.

Dr. Johnson addresses the major concerns of the *purchasers* of worksite health-promotion programs, such as reducing health costs, increasing productivity, and decreasing employee turnover. Developing more positive attitudes by employers toward their employees, increasing morale and creativity are some of the by-products of worksite programs that contribute to establishing a *positive* corporate culture, a hallmark of the AT&T "Total Life Concept (TLC)" programs.

Johannes Brengelmann describes the situation of worksite programs in Europe as substantively less well developed than in the U.S., albeit research on worksite stress has been particularly advanced in Sweden. Using stress-management programs as examples, several models currently in use in the Federal Republic of Germany are described.

A third major U.S. worksite health-promotion program, Control Data's "Staywell Program," is presented by William Jose to highlight issues of evaluation design and analysis, particularly with respect to organizational risk reduction and associated cost-savings analysis.

The second section of this book examines the "state of science" in four major components of worksite programs:

1. Smoking Cessation
2. Weight Control
3. Exercise
4. Stress Management.

Neil Grunberg reviews the evidence for health risks to the smoker (that are well documented) and to the nonsmoker being exposed to environmental tobacco at the worksite, which is a more controversial issue. The efficacy of the worksite smoking-cessation programs is reviewed, comparing different program models. The final section of the chapter reviews recent research findings on smoking and encourages incorporation of such findings (e.g., new information on the pharmokinetics of nicotine) to enhance the effectiveness of existing programs.

John Foreyt and Georganna Leavesley provide a succinct review of weight-control strategies, with particular attention to *maintenance* of weight loss. The history of worksite weight-reduction programs is reviewed, contrasting early program failures with recent more successful programs and the reasons for improved results. Recommendations for maintaining these results are offered.

The current research on the role of exercise in health maintenance is extensively reviewed by William Haskell, including cardiovascular disease, osteoporosis, diabetes, and obesity. Worksite programs are considered from the viewpoint of medical versus public health models, concluding that public health models that include more comprehensive environmental restructuring enhance long-term maintenance more effectively.

Dr. Kenneth Pelletier contrasts clinical and worksite stress-management programs, cautioning against uncritical extrapolation of data from the former to expectations from the latter. Guidelines for developing relevant worksite stress-management programs amenable to objective evaluation are discussed. This chapter was intended to be placed as Chapter 9, but for technical reasons, appears as Chapter 15.

The final section of this book focuses upon specific challenges to mounting effective and efficient worksite health-promotion programs. The chapter on program design and evaluation by Terry Mason addresses the importance of identifying program objectives, careful needs assessment, and objective program evaluation.

Legislative and policy issues associated with worksite health promotion (in the chapter by Margaret A. Hamburg) identifies many changes in worksite, demographics, and technology that form the basis for policy and legislative actions. Incentives to develop and maintain health-promotion activities are considered, including Federal tax incentives, technical assistance, and research programs.

Coordination between school, community, and worksite programs (in Chapter 11 by Richard Carleton) considers the interdependency and potential for positive interaction among programs. The complementary nature of the approaches (e.g., worksite vs. community) suggests that their effects may be additive rather than duplicative.

Recruitment and maintenance issues in worksite programs (in Chapter 12 by Kelly Brownell and William Jose) concerns encouraging individuals to enter programs and maintaining levels of participation. Factors affecting these variables are reviewed with guidelines for enhancing success in all these areas.

The chapter on cost-benefit and cost-effectiveness analysis by Jonathan E. Fielding points to the shortcomings of current cost-benefit and cost-effective analyses research in terms of study design, measurement problems, and so forth. Several examples of well-designed CBA and CEA studies are cited, with suggested guidelines for future studies.

The chapter on ethical issues by Marshall Becker raises important questions regarding the certainty of our knowledge base regarding health enhancement in terms of "risks vs. benefit" of health-promotion activity, and how one must eschew "overselling" the nature of the health risks involved.

As one reviews the range of topics and issues involved in worksite health promotion, several themes emerge:

1. "State-of-the-art" program design and evaluation (including cost-benefit and cost-effectiveness analyses) are essential ingredients to establish the necessary data base for developing, implementing, evaluating, refining, and maintaining effective programs.
2. One must be cautious in uncritically using data from "clinical" programs to support (or refute) worksite programs.
3. Changing behavior in health-enhancing directions is relatively straightforward; *maintaining* those changes is the real issue. The same strategies that work for *change* may not be the most appropriate ones for maintenance.
4. Although much worksite program development focuses on the individual, the social and environmental influences of the workplace can be powerful stimulants (or inhibitions) to health behavior change—and maintenance, influencing recruitment and program participation, as well.
5. Ethical, legislative, and policy concerns share many common goals and cautions.

6. There is a complementarity among public health-oriented programs (schools, community, worksite) that should be exploited by effective coordination by the involved parties.

The search for generic principles also includes the desirability of a generic model, in this case a ''health-enhancement planning model.'' Although ''health behavior'' in one form or another appears to be the generic issue in question, there are many interrelated factors (as previously noted) that must be considered in attempting to systematically organize the relevant variables into a cohesive planning model. Such a model should address three major sets of variables:

1. health behavior variables;
2. intra/interpersonal and environmental variables;
3. major situational variables.

As can be seen in Figure 1.1, health behavior can be subdivided into health behavior *development,* health behavior *change,* and health behavior *maintenance.*

Health behavior *development* refers to the actual *acquiring* of behaviors and/or skills related to particular practices that may be considered as either enhancing or detracting from one's health status. This component deals primarily with the developmental years, although in some cases this would also apply to adults, for example, in acquiring the habit of wearing seatbelts. Health behavior *change* is primarily concerned with the shifting from one behavior pattern to another or the cessation of health-destructive behavior patterns. Changing one's dietary patterns, stopping smoking, and shifting from a sedentary lifestyle to a more active lifestyle might be examples of such behaviors.

The third category, health behavior *maintenance,* has been purposely separated from the other two categories because of the chronicity and perhaps automaticity that we attempt to instill in this component of health behavior as compared to the more acute objectives involved in health behavior development and health behavior change. For example, the acquiring of a skill where there was none before or the changing of a habit from one form of behavior to another can identify the target behaviors desired and specify when they have been acquired at a given level of proficiency. The ''maintenance'' factor, however, involves indefinite periodic assessment of continued longevity of that pattern. It should also be recognized that the factors involved in maintaining a given behavior pattern may be quite different than those variables related to the acquiring or changing of that behavior pattern. (e.g., Hunt, Matarazzo, Weiss, & Gentry, 1979).

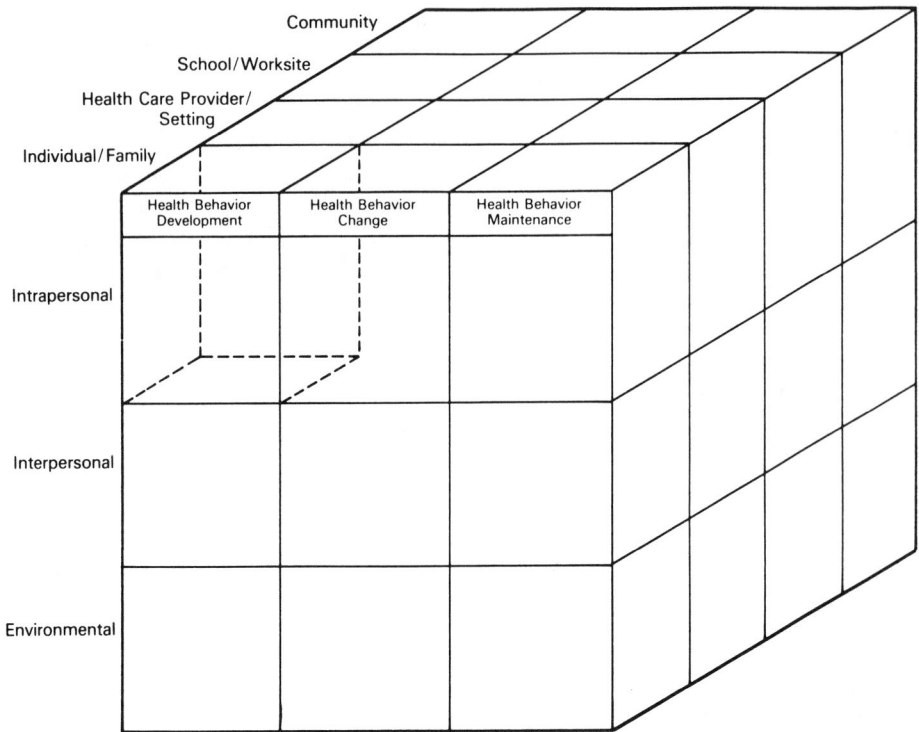

FIG. 1.1. Health-enhancement planning model.

There are three functional levels at which each of these health behavior variables must be considered. As noted in Figure 1.1, the "intrapersonal" level concerns itself with values, self-concept, personality characteristics, and other similar factors related to how individuals perceive themselves in terms of their personal effectiveness, competence, and worth. The "interpersonal" level refers to all forms of relationships with family, friends, colleagues, superiors, subordinates, and so forth. This also includes the quality of the relationships, whether they are competitive, cooperative, helping, social, loving, and so on. The "environmental" level includes all of the settings and related factors that affect the daily milieu of the individual. In addition to environments such as work, home, social, cultural, other factors such as climate, and the political/legal circumstances may also affect behavior in definable and predictable ways.

As one looks at the interplay between health behavior variables and functional levels, one recognizes that a variety of *situational* factors may also determine specific strategies in planning health-promotion programs. For example, one must consider such physical/personal venues as the individual in relation to his family and how this might affect intrapersonal values in terms of health behavior development or what interpersonal or environmental factors should be considered at the worksite in promoting health behavior change or health behavior maintenance.

Thus, it can be seen that through the various permutations of these three sets of variable, one can systematically explore the development of a comprehensive health-promotion strategy, utilizing all available resources and opportunities. Such a model would minimize some of the present confusion and fractionation of many of the current health-promotion efforts being carried on in various settings. It would also assist in identifying gaps in program planning and would provide a step-wise approach to health-promotion activities by targeting various aspects of the problem in a manner that would be consistent with resources available to the planning group.

In summary, it is obvious that the opportunities and challenges posed by worksite health-promotion activities are unique and potentially capable of significant impact on national health status and costs. Without doubt, progress is underway, as noted in the various chapters, but much more must be accomplished to place this area on a firm scientific footing.

In the best of the behavioral medicine tradition, the expertise of *many* disciplines, from economics to anthropology, health education to biomedicine, psychology to engineering, will be required to adequately address the multifaceted nature of the topic. The complexity of the issues involved are daunting, but I believe, resolvable, given the necessary time, expertise, money, and persistence. Scientific, federal, and private-sector resources must be joined to effectively meet the challenges posed in this volume; the necessary leadership must emerge from one of these groups if the potential synergy from such a "troika" is to be fully realized.

REFERENCES

Beary, C. A. (1981). *Good health for employees and reduced health care costs for industry.* Washington, DC: Health Insurance Institute.

Belloc, N. B., & Breslow, L. (1972). Relationship of physical health status and health practices. *Preventive Medicine, 1,* 409–421.

Doherty, K. (1988). Wellness at the worksite: Healthy returns. *Healthy Companies, 1,* 22–27.

Fries, J. F., & Crapo, L. M. (1981). *Vitality and aging.* San Francisco: Freeman Publishing.

Health Care Financing Administration. (1988). National health expenditures, 1986–2000. Washington, DC: Office of the Actuary, Division of National Cost Estimates.

Hunt, W. A., Matarazzo, J. D., Weiss, S. M., & Gentry, W. D. (1979). Associative learning, habit and health behavior. *Journal of Behavioral Medicine, 2*(2), 111–124.

Keys, A., Taylor, H. L., Blackburn, H., Brozek, J., Anderson, J. T., & Simonson, E. (1971). Mortality and coronary heart disease among men studied for 23 years. *Archives of Internal Medicine, 128,* 201–214.

Rosenman, R. H., Brand, R. J., Jenkins, C. D., Friedman, M., Straus, R., & Wurm, M. (1975). Coronary heart disease in the Western Collaborative Group Study: Final follow-up experience of 8½ years. *Journal of the American Medical Association, 233,* 872–877.

World Health Organization. (1948). *Constitution of the World Health Organization.* Geneva: WHO Basic Documents.

WORKSITE HEALTH PROMOTION STRATEGIES AND PROGRAMS

2 The Challenges of Work-place Health Promotion

Jonathan E. Fielding
Schools of Public Health and Medicine
University of California, Los Angeles
and
Johnson & Johnson Health Management, Inc.

The continued strengthening of causal links between deleterious health and the major causes of death and disability in the working age population suggests significant opportunities for employers to invest in improving employee health. Most employers also pay the majority of health-benefits costs for employees, and often for their dependents and retirees. The high benefit-plan costs of treating many preventable diseases is one of the rationales for employers to invest in risk-reduction efforts.

Organized health-promotion efforts at the work place, though promising, have only achieved significant market penetration within the 1980s, and evaluations to date are limited. Thus, examination of the current status of these programs and the major challenges faced in determining their proper role is timely.

Three topics will be covered in this chapter:

(1) the prevalence of health-promotion programs among employers in the United States;
(2) types of evaluation approaches and results of several comprehensive employer-sponsored health-promotion programs; and
(3) a selective list of some of the major scientific challenges in this emerging field.

HEALTH-PROMOTION PROGRAMS IN U.S. WORK SITES

A limited number of systematic studies of the prevalence and nature of work-site health-promotion programs have been conducted.

A study of health-promotion programs at California work sites was carried out by Fielding and Breslow (1983), while Davis, Rosenberg, Iverson, Vernon, and Bauer (1984) conducted a similar study of Colorado work sites. The Departments of Health of Minnesota, (1982), Rhode Island, (1982) and Texas (Fellows, Gottlieb, & McAlister, 1988) each have surveyed the extent of health-promotion programs in industry. These studies have all provided some data on program frequency and content. However, because the studies were geographically limited, conducted at different times, and used different instruments, the results could not be easily combined or extrapolated to provide national estimates.

To meet the need for a baseline national study, the Office of Disease Prevention and Health Promotion in the Department of Health and Human Services contracted with U.S. Corporate Health Management (Health Promotion Division; now merged into Johnson & Johnson Health Management, Inc.) to carry out a survey on health promotion at the work site that would provide national estimates of the prevalence of work-site health-promotion activities.

Analysis of the frequency data from that survey has been published (Fielding & Piserchia, 1989) and provides a national snapshot of health-promotion activities based on a random-sample survey of work sites with 50 or more employees ($n = 1358$; see Table 2.1).

At least one health-promotion activity was reported by 65.5% of the work sites surveyed. Although activities were common in work sites of all sizes, the frequency of health-promotion activity increased directly with work-site size. Work-site respondents were queried on whether they had any of nine specific types of health-promotion activities. Smoking cessation and control was the most prevalent activity (35.6%), followed by health-risk assessment (29.5%), back-problem prevention or care (28.5%), stress management (26.6%), exercise and fitness activities (22.1%), off-the-job accident prevention (19.8%), nutrition education (16.8%), blood-pressure treatment and control (16.5%), and weight control (14.7%; Fielding & Piserchia, 1989).

The range of initiatives reported by the work sites included information, policy, individual counseling, group classes or workshops, and special promotional events.

Health-education information through newsletters, brochures, bulletin boards, or pay-envelope stuffers was provided at all of the work sites with activities. Individual or group education and risk-reduction activities were found with lower frequency, depending on the specific activity, work-site size, and industry type.

An example of a work-site health-promotion policy is a defined no-smoking policy, which was found at 76.5% of the work sites with smoking-cessation activities. A seat-belt policy for company vehicles was in effect at 61.5% of the work sites with off-the-job accident-prevention activities.

TABLE 2.1
Demographic Table of First National Work Site
Health-Promotion Survey

Work-site Size	
Size	*Percentage*
<100	50.2
100–249	30.3
250–749	12.9
>750	6.6
	100.0

Region	
Area	*Percentage*
Northeast	23.6
North Central	25.4
South	32.6
West	18.4
	100.0

Industry Type	
Industry	*Percentage*
Manufacturing	29.4
Wholesale/Retail	15.7
Utilities/Transportation/Communication	3.6
Financial/Real Estate/Insurance	6.2
Services	37.5
Other	7.6
	100.0

Source: Adapted from Fielding and Piserchia (1989).

Work sites reporting a high stress level among employees were more likely to provide stress-management activities, and work sites that required heavy physical work were more likely to present programs on prevention and care of back problems.

Incentives such as rewards, cash bonuses, time off, or recognition were frequently provided for participating in health-promotion activities and/or for health behavior change. At the majority of the work sites providing individual or group activities, the company either offered these gratis to employees and/or provided employees with time off work to participate (Fielding, 1987).

Almost all of the employers with any health-promotion activities felt that there were benefits to the activities and that the benefits outweighed the cost to

TABLE 2.2
First National Work-site Health Promotion Survey:
Reasons for Offering Health-Promotion Activities
(Percentage of work sites with activities)

Reason*	Percentage
Improve employee morale	9.7
Reduce health-insurance costs	10.0
Improve employee health	28.0
Reduce disability claims and lost time	5.2
Decrease hospital / medical utilization	0.6
Increase output/productivity/quality	10.3
Reduce accidents on the job	5.1
Reduce accidents off the job	0.8
Improve corporate image	1.4
Management wanted it	17.6
Employees wanted it	9.3
Unions wanted it	0.5
Other companies were doing it	1.4
Other reasons	39.6

Source: Fielding (1987).
*More than one response is possible.

the company. However, only 25% of the work sites indicated that they had a formal evaluation of the activities they were sponsoring (Fielding, 1987).

The most compelling reason for sponsoring health-promotion activities was to improve employee health, which was listed by 28% of the work sites with activities. Other commonly cited reasons were that management wanted the activities; that it was required by federal, state or local law; or that it was thought to be "the right thing to do" (see Table 2.2; Fielding, 1987).

Management's role and level of interest were judged to be important determinants of program success by most respondents. The chief executive officer was considered to be committed to health-promotion activities at 49% of the work sites with activities, while the employees were thought to be very committed at only 13.9%, and the unions at 4.9%. Commitment by management is reflected by the financial support with the majority of employers paying for individual counseling, for groups and workshops, and for screenings and physical examinations. Most types of activities (except health-risk screenings and examinations) were held predominantly at the work site and on company time. All permanent employees were eligible for health-promotion activities at 85.6% of the work sites, employee dependents at 30.1%, and retirees at 30.4%. Money for health-promotion activities was a separate cost center at 35.1% of the work sites with activities. At other work sites, the funding came from the medical department (17.5%), personnel or benefits (55.3%), or other corporate budget (33.8%; see Table 2.3; Fielding, 1987).

This initial national survey of employer-sponsored health-promotion activities, coupled with the results of smaller state surveys, demonstrates that work-site health-promotion programs are common, appear to be growing in frequency and scope, and tend to be more comprehensive in larger work sites. The paucity of reported formal evaluations suggests that many employers are willing to sponsor such activities without an objective way to demonstrate benefits.

Despite the high frequency of work-site activities, few such programs would meet most definitions of a comprehensive health-promotion program. Even less common are evaluations of these comprehensive programs in academic-quality journals. Nonetheless, several different types of evaluation activities have been undertaken in connection with such programs. For example, at least two corporations have been the subject of published studies comparing baseline characteristics of program participants and nonparticipants. The Control Data Corporation, in collaboration with an actuarial firm, has published primarily on the excess costs of bad health habits and related physiological and biochemical health indices under their health-benefit plans. Dichotomizing each employee health behavior analyzed into three risk categories, the excess claims costs of the high-risk group over the low-risk group were 11% for weight, 114% for exercise, 118% for smoking, 11% for hypertension, and 113% for seat-belt use. High-risk alcohol users (based on self-report of intake) had claims costs 9% less than the low-risk users, which may be due to underreporting of high use or to the fact that high-risk alcohol users terminate employment with the company at higher rates than do their low-risk counterparts (Brink, 1987).

A series of articles based on the Tenneco comprehensive health-promotion program in which fitness appears to be the most important component has queried the relationship of participation in the fitness program to productivity and employer cost variables. These provocative studies have focused on base-

TABLE 2.3
First National Work-site Health Promotion Survey:
Sources of Health-Promotion Activity Funding
(Percentage of work sites with activites)

Source of Funding	Percentage
Budget or money specifically allocated	35.1
Funding from medical department budget	17.5
Funding from personnel or benefits budget	55.2
Funding from other corporate budget	33.8
Funding from insured benefits	16.1
Funding from voluntary health organization funds	8.7
Funding from any other source	10.1

Source: Fielding (1987).

AGE

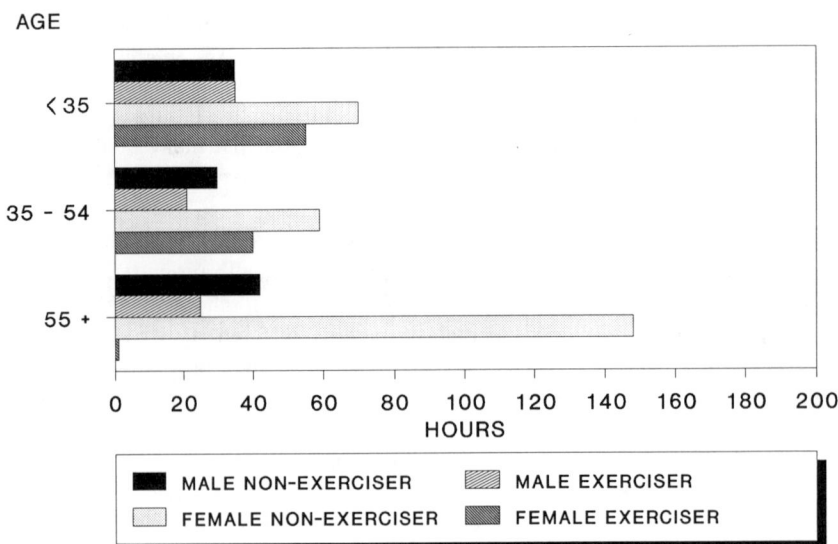

FIG. 2.1. Tenneco, Inc. Health and Fitness Program Analysis Data: Relationship of age and absenteeism to exercise status. Source: Baun et al., 1986.

line differences between participants and nonparticipants, and therefore no implication can be drawn regarding program effects. Strong white-collar employee adherence to participation in the on-site formal exercise program was strongly associated with above-average (job) performance based on personnel assessments ($p < .01$; Bernacki & Baun, 1984). During a four-year period, turnover was significantly higher among nonexercisers than exercisers, after taking into account age, gender, general job category, and duration of employment (Tsai, Baun, & Bernacki, 1987). Compared to nonexercisers in the corporate program, female but not male exercisers had significantly lower absenteeism [47 vs. 69 hours ($p < .05$); see Fig. 2.1]. Total health-care costs among exercisers tended to be lower but not significantly so, but ambulatory health-care costs were significantly lower for exercisers (Baun, Bernacki, & Tsai, 1986).

Examples of evaluation of the impact of comprehensive health-promotion programs come from the experience of AT&T and Johnson & Johnson. AT&T has published evaluation data primarily dealing with changes in health behaviors and health risks associated with the implementation of its TLC comprehensive health-promotion program (Bellingham, Johnson, & McCauley, 1985). In summary, a report on this program piloted at AT&T Communications used a combination of true experimental and quasi-experimental design. The authors identified a study group composed of two sites, one in which all employees were offered the program and the other where it was offered to a random sample of employees. Members of the study group were given an initial health-risk appraisal and offered health-education modules during the following 12 months. The study groups were compared with two control groups, one given only the health-risk appraisal and the other given nothing. Based on a 12-month health-risk appraisal follow-up and ANCOVA analysis for 10 key dependent variables through a second health-risk appraisal, significant ($p < .05$) differences in scores between individuals in the two study sites versus the controls were identified. However, changes in scores were limited to Type A behavior and body weight at one study site, and diastolic blood pressure and serum cholesterol at the other. Based on health-risk appraisal calculations, the risk of heart attack and overall risk of dying in the next 10 years were significantly reduced for the total study group (see Table 2.4; Spilman, Goetz, Schultz, Bellingham, & Johnson, 1986).

Significant improvements in self-reported productivity measures were observed in the study groups compared to the controls, but without adjustment for demographic differences. No consistent pattern of change was observed in psychological well-being, work enthusiasm, satisfaction with working condi-

TABLE 2.4
AT&T TLC Program Evaluation: Significant Differences
Between Study Groups and Control Groups
(ANCOVA)

	Kansas City	Bedminster
Individual		
Systolic blood pressure	No	No
Diastolic blood pressure	No	No*
Serum cholesterol	No	No*
High-density lipoprotein	No	No
Type A behavior pattern	Yes*	No
Alcohol consumption	No	No
Body weight	Yes*	No
Weekly physical exercise	No	No
Health age/age	No	No
Health age/attainable	No	No
Group		
Total mortality/average	Yes*	No
Total mortality/attainable	No*	Yes*
Heart attack morbidity/average	No	No
Heart attack morbidity/attainable	Yes*	Yes*
Cancer morbidity/average	No	No
Cancer morbidity/attainable	Yes*	No

*Statistically significant at 0.05 or less.
Source: Spilman et al., 1986.

tions, work pressure, and satisfaction with the work place. Interpretation of results is limited by low participation (28%), marked differences in demographic differences in the no-intervention control group, and reliance on self-report for some variables for which reliability may be questionable.

Johnson & Johnson's LIVE FOR LIFE® program is discussed in greater detail because it has been the subject of more published evaluation results in peer-reviewed journals. The principal goals of LIVE FOR LIFE are to provide the opportunity and encouragement for Johnson & Johnson employees to become the healthiest in the world and to control the corporate costs of employee ill health (Bly, Jones, & Richardson, 1986). LIVE FOR LIFE is a comprehensive effort aimed at helping all employees at a work site to improve and maintain their health by establishing good health habits.

Although the program is modified to suit the particular work site's special needs and demographics of the employed population, standard components include: (1) a health screening which includes a questionnaire on health habits and attitudes, a series of health measurements (e.g., blood pressure, body fat, blood lipids, height, weight, and estimated maximal oxygen uptake), and health-education counseling by a specially trained nurse; (2) communications programs (newsletters, health fairs, contests, posters, etc.) designed to improve awareness and to encourage high levels of employee participation; (3) a lifestyle seminar to introduce employees to the program; and (4) a variety of behavior-change-oriented action programs on subjects such as smoking cessation, weight control, stress management, blood-pressure reduction, nutrition, and fitness. These voluntary programs are generally free of charge to employees, and most activities occur at the work site. Strong efforts are made to change the work environment, with attention to smoking policies, offering nutritious food, and providing facilities for exercise. Most activities occur on the employee's own time.

LIVE FOR LIFE programs were initiated in 1979 and phased in at Johnson & Johnson companies, covering 31,200 employees in 1987. The first evaluation involved a quasi-experimental design in seven Johnson & Johnson plants in Pennsylvania and New Jersey. Selected characteristics of LIVE FOR LIFE and control-group volunteers are summarized in Table 2.5. Although some statistically significant differences exist, the three treatment and four control companies have generally similar health habits, overall, with the exception of smoking, in which 26.9% of employees smoked in treatment sites, versus 35% for control sites. Overall, the average age was 34.9 for men and 34.1 for women (Wilbur, Hartwell, & Piserchia, 1986).

Two-year results for the LIVE FOR LIFE employees ($n = 1,399$) who completed two health screenings between 1980 and 1982 were compared with the group of control-site employees ($n = 748$) who completed two health screenings during the same period. LIVE FOR LIFE employees had a significantly higher smoking-cessation rate (23% versus 17%), based on self-report

TABLE 2.5
Johnson & Johnson LIVE FOR LIFE Epidemiologic Study:
Selected Baseline Characteristics of Health Screening
Volunteers by Treatment and Gender

	Female		Male	
	Control	Treatment	Control	Treatment
Number of volunteers	751	1272	518	971
Volunteer rate (%)	73.5	79.3	77.3	77.7
Mean age (in years)	34.8	33.7	36.1	34.3*
Age distribution (%)				
18–24	24.3	22.4	8.1	11.1
25–34	35.1	41.0	43.6	48.2
35–44	14.7	16.8*	28.8	25.9
45–54	19.0	12.6	14.1	10.5
55	6.9	7.2	5.4	4.3
Total	100.0	100.0	100.0	100.0
Ethnicity (%) minority	9.8	16.4*	10.0	15.1*
College graduates (%)	11.6	20.7*	58.4	61.8
Smoking distribution (%)				
Never smoker	40.5	49.7	40.7	50.1
Former smoker	17.4	19.6	32.3	26.8
Current smoker	42.1	30.7*	27.0	23.1*
Total	100.0	100.0	100.0	100.0
Mean cigarettes per day (current smokers)	18.3	18.5	23.3	21.9
Mean total ethanol per week (oz)	2.4	23.8	4.6	4.0
Mean systolic blood pressure (mm Hg)	116.3	114.4	126.7	124.8*
Mean diastolic blood pressure (mm Hg)	75.1	73.3*	81.5	80.8
Hypertensive (% w/SBP of 140 or DBP of 90)	8.4	6.3	22.1	16.5*
Mean % above ideal weight[a]	15.1	11.7*	16.9	15.9
Mean total activity (kcal / kg / day)[b]	34.4	34.8	36.9	36.9
Mean general well being[c]	76.8	77.2	82.0	81.5
Mean Framingham Type A behavior scale[d]	6.1	6.0	5.3	5.3
Mean number of self-reported sick days	6.1	5.6	2.7	3.8*
Mean job satisfaction with growth opportunities	59.5	61.0	63.8	62.9
Mean satisfaction with supervision	66.9	65.7	64.3	63.1
Mean satisfaction with working conditions	71.3	71.6	70.9	62.9

[a]Ideal weight is 5 pounds greater than the values in the Metropolitan Life Insurance tables.

[b]Estimated from a previously validated, seven-day physical activity recall interview (Blair et al., 1980).

[c]A scale based on a series of 18 questions developed from the National Center for Health Statistics (Frazio, 1977).

[d]A scale based on a series of 10 questions developed for the Framingham Study (Haynes, Levine, & Scotch, 1978).

*Difference between treatment and control with gender group is statistically significant at $p = .01$.

Source: Wilbur et al., 1986.

and verified by blood thiocyanate. The cessation-rate difference was even more pronounced for employees at high risk for coronary heart disease (32% vs. 13%; Shipley, Orleans, Wilbur, Piserchia, & McFadden, 1988). LIVE FOR LIFE work-site employees lost significantly more weight or gained less weight than did control employees. Overweight employees in LIVE FOR LIFE sites lost an average of 1.1 pounds, while overweight controls gained an average of 0.5 pounds. Note that these results relate to *all* employees who took two health screenings and were overweight, regardless of whether they participated in any formal weight-reduction, nutrition, or exercise programs (Jeffery, 1983).

Assessment of changes in exercise over the two-year period came from 1,399 employees at LIVE FOR LIFE companies and 748 employees at control companies, 95.2% and 94.3% respectively of all employees for whom data was available at baseline and two years. Physical fitness was directly assessed by estimating maximum oxygen uptake from a submaximal test (75% of projected maximum heart rate) on a bicycle ergometer. Three questionnaire-based physical activity self-report techniques were utilized to determine the nature and level of physical activity. Participants were asked their level of physical activity compared to others based on a seven-point scale. A second question asked if they regularly performed activities which equated to energy expenditure of at least 1,000 kcal/wk for a 70 kg individual, the minimum level recommended by many experts. A third approach was a seven-day recall of physical activity (Blair, Piserchia, Wilbur, & Crowder, 1986).

Almost 20% of women and 30% of men in LIVE FOR LIFE companies initiated regular vigorous exercise during the two-year period compared with 7% and 19% for women and men in control companies. Maximum oxygen consumption increased 8.4% in year 1 and 10.5% in year 2, compared to baseline in LIVE FOR LIFE employees, versus 1.5% and 4.7% in years 1 and 2 respectively for the control group. Sociodemographic variables did not influence maximum oxygen uptake, based on the analysis using the general linear model procedures with interaction terms for group program status and many sociodemographic variables. Attrition bias was excluded as a source of intergroup differences. Larger changes in maximum oxygen uptake capacity were significantly correlated with reduction in body weight, percent body fat, and systolic blood pressure, but not diastolic blood pressure or blood lipid components (Blair et al., 1986).

Program impact on health-benefit-related costs borne by Johnson & Johnson was also assessed. Employee medical claims paid by Johnson & Johnson during the period 1979–1983 were used for analysis based on edited claims tapes. All employees were subject to the same benefit plan administered by the same health-insurance carriers. Johnson & Johnson companies were divided into three groups based on the length of time that LIVE FOR LIFE had been in

TABLE 2.6
Johnson & Johnson LIVE FOR LIFE Economic Study:
Medical Claims, 1979 to 1983*

Year	Unadjusted Means, Group			Adjusted Means, Group		
	1	2	3	1	2	3
Inpatient Costs per Employee ($)						
1979	89	78	87	. . .	. . .	. . .
1980	102	131	111	110	144**	100
1981	158	164	150	167	171	149
1982	177	195	224	190	207	229
1983	233	212	424	265***	258***	403

*Adjusted means for 1980 to 1983 include adjustments to raw means based on age, gender, job class, New Jersey status, and baseline (1979) differences. Standard errors of estimate related to the adjusted means ranged from $7.16 in 1980 to $15.75 in 1983 for inpatient costs. For days of hospitalization, the standard errors ranged from 28.0 days per 1,000 in 1980 to 54.6 in 1983. For admissions, the standard errors ranged from 3.2 per 1,000 in 1980 to 5.6 in 1983.
**$p = .01$ to .05.
***$p = .001$ to .005. Comparisons are made to group 3.
Source: Bly et al., 1986.

operation: (1) > 30 months as of December 1983; (2) 18–30 months as of December 1983; or (3) no LIVE FOR LIFE program as of December 1983. Employees who had been continuously employed in one of these three groups of companies from 1979 to 1983 were selected for analysis, with group sizes from about 3,000 to over 5,000. Analysis of covariance was employed to adjust for differences among groups with respect to gender, age, job class, baseline utilization levels and costs, and location (Bly et al., 1986).

The most striking finding was for inpatient costs. The average employee in LIVE FOR LIFE companies experienced significantly slower increases than did his control group (group 3) counterpart, with intergroup differences due to inpatient costs. Mean annual overall cost increases, in constant dollars, were $43 and $42 for the two LIVE FOR LIFE groups versus $76 for the control group. LIVE FOR LIFE groups also experienced lower rates of increase in hospital days and admissions to acute-care hospitals (see Table 2.6; Bly et al., 1986).

Analysis of possible LIVE FOR LIFE program impacts on absenteeism compared employees at four LIVE FOR LIFE sites and five control sites. Analysis was performed for the group of employees who completed health and lifestyle questionnaires during 1979 and again in 1981, and over a two-year period LIVE FOR LIFE was associated with a significant ($p < 0.01$) reduction in absenteeism for wage employees but not for salaried employees (Jones et al., 1990).

WORK-SITE CHALLENGES

The scientific base related to work-site health-promotion programs is thin, and there are a spate of basic issues that are more often subjected to debate than peer-reviewed studies. The following ten, easily identified from an admittedly larger universe, should become priority subjects for research:

1. *Design issues.* Implementing the best evaluation design, especially if the unit of analysis is the work site, can be difficult to achieve. Randomization of work sites is rarely possible, requiring recourse to quasi-experimental designs. Randomization of individual employees to multiple conditions is more often feasible, although the use of a no-intervention control is frequently difficult to sell to employers.

The methodological purity and meticulous data collection procedures necessary for a high-quality intervention study are often difficult to achieve. Many work sites have difficulty seeing how the limited benefit to them can justify the administrative work, interference with usual procedures, and extra demands on their employees.

2. *Environmental influences on health.* Each work site has a distinct culture, which influences the behavior, including health-relevant behavior, of employees. How can a "corporate culture" be defined in a reproducible manner? What are its dimensions? What disciplines should best be combined in fashioning a definition? Were this base clearly established, we could move to studies associating specific aspects of corporate culture and efforts to modify it with health habits and other dependent variables.

3. *Impact of policy changes.* Work sites set policies which can influence employee health directly or indirectly. Policies that ensure safety on the job or attempt to protect against burnout (e.g., require annual vacations or job rotation) may reduce the incidence of health problems directly related to the job. Employers may also set policies which can or could influence health habits. The majority of employers (57.5%) with more than 750 employees have a smoking policy which specifies where and under what circumstances smoking is permitted during working hours (Fielding, 1987). Some employers pay employees for time involved in health-promotion activities during the work day, such as periodic health screening or participation in exercise on site or off site. What impact do these policies have on disease frequency and severity, as well as on intermediate measures, such as fitness level and smoking status? What is the slope of any observed change? Do size and longevity of effect depend on mode of introduction or source of policy determination, nature of work force, and/or type of work? How reproducible are effects of specific policies in different work sites? Do enforcement mechanisms strongly affect compliance?

24

4. *Recruitment strategies.* When the target of a work-site health-improvement program is the entire population, or all of those with a particular ameliorable health problem, participation is crucial to program success. Many different recruitment strategies have been employed to secure high participation rates, including a broad array of media to communicate the benefits of participation: personal invitations; well-publicized participation by key management personnel; and both monetary and nonmonetary incentives. How well do alternative strategies work, alone or in combination? Do specific strategies work best for specific programs (e.g., a videotape introduction for health screening, showing workers going through the parts of the exam; personal invitation from a co-worker to attend a weight-management class; and indications of a willingness to help the participant to get through the rough spots)?

5. *Different target groups.* Those at highest risk may be the hardest to reach. What approaches are efficacious in both obtaining involvement and in predicting success? Do high-risk individuals need special programs, or do they do best if they receive standard interventions? To what degree do "standard" approaches need to be customized to appeal to sub-segments of the population, defined by age, ethnicity, or cultural heritage? Do certain customizations achieve reproducible results (with respect to knowledge, attitudes, participation, recidivism) in different work sites?

6. *Productivity measurements.* Current state of knowledge regarding reliable and valid measurement of productivity is rudimentary, with the exception of a limited number of jobs involving repetitive tasks with easily quantifiable quality-assurance measures. How to best measure productivity and whether health-promotion program presence can be related to changes in productivity are as yet unknown, except for such discrete components as health-benefit costs and absenteeism. One important question is whether there are reproducible correlations between responses to standard questionnaires that measure morale, job satisfaction, satisfaction with employer, etc., and productivity defined in terms of variables of importance to particular employers, such as reduced turnover, better supervisor evaluations, and/or faster rates of promotion.

7. *Impact of health habits on employer health costs.* Most employers pay the majority of the costs for covered health care of employees, and usually also of dependents. Part of the economic argument for employer attitudinal and financial support of health-promotion programs is that poor health habits cost more. Yet few studies have quantitatively linked the frequency and degree of certain health habits with employer-paid costs for health-care services.

Even fewer studies have related the introduction of health-promotion programs at the work site to changes in spending for these services. If there are changes, will these be among the insured who tend to use few services, or those 2–3% of claimants who account for 40–60% of the total costs (Fielding, 1984).

8. *Smaller work sites.* Most of the literature reports treat work sites of 500–5,000 employees, yet the majority of the work force is employed in work sites with 50 or fewer employees. What are effective and cost-effective strategies for smaller work sites, and do they differ from ones easily employed in larger settings? Significant numbers of such work sites may be required for purposes of any formative or summative evaluation to obtain sufficient sample sizes.

9. *Program renewal.* A program that achieves its objective during a particular year may not work as well several years hence. Health-promotion programs, like any product or service, have a life cycle and require reconceptualization to respond to changing market conditions. What is a reproducible process for taking the pulse of the "market" for an intervention program and making minor or major changes that achieve results in line with expectations and with previous experience?

10. *Is there a "healthy employer?"* Can the concept of "healthy employer" be operationalized? Although currently there is no accepted definition of this concept, its core would be an accepted culture supportive of health improvement and maintenance in its employees, and possibly in dependents as well. Does it follow that an employer supportive of smoking cessation and on-the-job injury prevention will, naturally, also be supportive of stress management, of improved nutrition, or of increased physical activity? To what degree is this concept a direct and temporally dependent function of the priorities and perspective of a few top managers? Can the infrastructure for a healthy company be constructed so that its endurance is assured, regardless of reorganizations, changes in senior management, and economic cycles?

The list of issues and related challenges can easily be extended. These ten selected issues exemplify the opportunities for high-quality research in this relatively uncharted, but growing field.

FUTURE TRENDS

At least four major trends are likely to extend into the future. First, both the prevalence and types of health-promotion programs sponsored by employers are likely to continue to grow, with the possible exception of periods of economic recession. Second, employers are likely to become more sophisticated managers and purchasers of health-promotion services, asking more questions about program efficiency and effectiveness, but rarely willing to fund academic-quality evaluation. Third, multidisciplinary evaluation teams, which may include epidemiologists, behavioral-medicine experts, biostatisticians, economists, sociologists, and possibly other disciplines, will increasingly use work sites to test hypotheses about health-improvement interventions, and will be more likely to receive support from federal and other grant sources for research and evaluation efforts using employee populations.

Finally, during the 1990s, researchers will devote substantial efforts to develop models testing the interrelationships among health status, productivity measures, and employee attitudes, as well as characteristics of the work site that influence these variables. The research agenda may appear daunting, but the stakes are not only maximizing health status but competitiveness in a global economy, a high-priority interest shared by the public and private sectors.

REFERENCES

Baun, W. B., Bernacki, E. J., & Tsai, S. P. (1986). A preliminary investigation: Effect of a corporate fitness program on absenteeism and health care cost. *Journal of Occupational Medicine, 28,* 18–22.

Bellingham, R., Johnson, D., & McCauley, M. (1985). The AT&T Communications Total Life Concept. *Corporate Commentary, 1,* 1–13.

Bernacki, E. J., & Baun, W. B. (1984). The relationship of job performance to exercise adherence in a corporate fitness program. *Journal of Occupational Medicine, 26,* 529–531.

Blair, S. N., Haskell, W. L., Ho, P., Paffenbarger, R. S., Jr., Vranizan, K. M., Farquhar, J. W., & Woods, P. D. (1980). *Measuring physical activity by a seven-day recall method. Crisis in the public sector* (program and abstracts, p. 232). Waashington, DC: American Public Health Association.

Blair, S. N., Piserchia, P. V., Wilbur, C. S., & Crowder, J. H. (1986). Public health intervention model for work-site health promotion. *Journal of the American Medical Association, 255,* 921–926.

Bly, J. L., Jones, R. C., & Richardson, J. E. (1986). Impact of worksite health promotion on health care costs and utilization. *Journal of the American Medical Association, 256,* 3235–3240.

Brink, S. D. (1987). *Health risks and behavior: The impact on medical costs.* Milwaukee: Milliman & Robertson, Inc.

Davis, J. F., Rosenberg, K., Iverson, D. C., Vernon, T. M., & Bauer, J. (1984). Worksite health promotion in Colorado. *Public Health Reports, 99,* 538–543.

Fellows, J., Gottlieb, N. H., & McAlister, A. L. (1988). Employee health promotion: Organization correlates and community resources. *High Level Wellness, 12,* 5–15.

Fielding, J. E. (1984). *Corporate Health Management.* Reading, MA: Addison-Wesley.

Fielding, J. E. (1987). *National Worksite Health Promotion Activity Survey.* Unpublished data.

Fielding, J. E., & Breslow, L. (1983). Health promotion programs sponsored by California employers. *American Journal of Public Health, 73,* 538–542.

Fielding, J. E., & Piserchia, P. V. (1989). Frequency of worksite health promotion activities. *American Journal of Public Health, 78,* 16–20.

Frazio, A. F. (1977). A concurrent validation study of the NCHS General Well-Being Schedule (DHEW Publication No. HRA 78–1347). *Data Evaluation and Methods Research, Series 2, No. 73,* 1–52. Washington, DC: U.S. Government Printing Office.

Haynes, S. G., Levine, S., & Scotch, V. (1978). The relationship of psychosocial factors to coronary heart disease in the Framingham Study: I. Methods and risk factors. *American Journal of Epidemiology, 107,* 362–383.

Jeffery, R. W. (1983). *The effect of the LIVE FOR LIFE Program on employee weight status. Johnson & Johnson internal review.* Unpublished manuscript, New Brunswick, NJ.

Jones, R. C., Bly, J. L., & Richardson, J. E. (1990). A study of worksite health promotion and absenteeism. Evaluation of the LIVE FOR LIFE Program. *Journal of Occupational Medicine, 32,* 95–99.

Minnesota Department of Health. (1982). *Workplace health promotion survey.* Minneapolis: Minnesota Department of Health.

Rhode Island Department of Health. (1982). *The role of Rhode Island industry in health promotion.* Providence: Rhode Island Department of Health.

Shipley, R. H., Orleans, C. T., Wilbur, C. S., Piserchia, P. V., & McFadden, D. W. (1988). Effect of the Johnson & Johnson LIVE FOR LIFE Program on employee smoking. *Preventive Medicine, 17,* 25–34.

Spilman, M. A., Goetz, A., Schultz, J., Bellingham, R., & Johnson, D. (1986). Effects of a corporate health promotion program. *Journal of Occupational Medicine, 28,* 285–289.

Tsai, S. P., Baun, W. B., & Bernacki, E. J. (1987). Relationship of employee turnover to exercise adherence in a corporate fitness program. *Journal of Occupational Medicine, 29,* 572–575.

Wilbur, C. S., Hartwell, T. D., & Piserchia, P. V. (1986). The Johnson & Johnson LIVE FOR LIFE Program: Its organization and evaluation plan. In M. F. Cataldo & T. J. Coates (Eds.), *Health and industry: A behavioral medicine perspective* (pp. 338–350). New York: John Wiley.

3 Health Promotion: The Challenge to Industry

D. R. Johnson
AT&T

The topic of this paper is timely as there are currently many issues surfacing during this accelerated growth period within the field of health promotion.

According to a National Survey of Worksite Health Promotion (1987) sponsored by the U.S. Public Health Service in 1985, 65% of U.S. companies having 50 or more employees support some form of in-house health-promotion program. That reflects a dramatic increase of approximately 20% over the last few years. The basis for that growth appears to be related to three factors.

The first is the drive to manage benefit-related medical expenses. The second is the premise that healthier employees help to build healthier businesses. And the third relates to the desire to develop a positive corporate culture.

The first reason—benefit-related cost management—is probably providing the greatest incentive. Indeed, with health-care costs rising at an 8 to 10% rate annually [the total for 1986 reaching an unprecedented $465.4 billion (U.S. Industrial Outlook, 1987)] companies, both large and small, are taking a hard look at what they can do in response to the cost dilemma as more than one-fourth of the total bill is subsidized by the nation's employers (Opatz, 1986).

What, then, are the alternatives? Cut benefits. Negotiate more favorable insurance contracts. Insist on second opinions for elective surgery. Encourage the use of HMOs or alternative plans. Promote good health (Herzlinger, 1985a, 1985b, 1986).

Of all the alternatives cited, the last, achieved in part through worksite health-promotion programs, has the greatest long-term potential to contain medical expenses. In addition to the cost-management effect, these programs also have the potential to achieve outcomes such as improved attitude and morale, which are usually not aligned with the other cost-cutting strategies cited.

The philosophy behind health-promotion programs is simple—to encourage employees to develop lifestyles that will minimize health risks while at the same time decrease the likelihood of ever having to use medical benefits. In other words, healthier lifestyles reduce the risk of illness, disability, and premature death, which enables the business to reduce its medical and life-insurance expenses.

At AT&T, for instance, about 1,000 employees have quit smoking after participating in our smoking-cessation program. Because of that intervention, those 1,000 employees will visit a doctor much less, have fewer total sick days when faced with a serious illness, and enjoy a heightened quality of life. The average avoided costs for the corporation as a result of this program is estimated at approximately $4,000 per employee per year—that's $4 million per year just for those 1,000 employees (Kristein, 1983).

Thus, wellness programs can have a *direct* impact on the bottom line. But just as important is the *indirect* effect that comes by way of increased productivity, increased creativity, and increased energy and stamina that healthier employees generally bring to their work.

Speculate for a moment. If, on the one hand, there is an employee who smokes, is overweight, and mismanages hypertension but, on the other, is an employee who exercises, has sound dietary habits, and has learned how to manage stress—who do you think is going to be more productive, more creative, and more vibrant? Who do you think is going to do a better job for the business? This is, to some degree, speculative, but today, in 1987, it is less speculative than in the past as researchers are working hard to provide empirical evidence linked to increased productivity.

Finally, companies are becoming more involved with health-promotion programs because of the effect such activity has on building a positive corporate culture. Health-promotion programs incorporate good human-resource management into the business plan, and, in an attempt to *change* a corporate culture, such programs become part of the overall strategic thrust.

Why? There are any number of reasons, not the least of which has to do with basic principles of competition. Today's employees don't salute the company flag like they used to. They're more mobile. They seek a lot more than a paycheck. And the more talented an individual is, the more likely it is that he or she will look beyond compensation when deciding where to work.

Furthermore, health-promotion programs demonstrate a genuine interest in people. They show that the company stands behind its employees and is willing to go the extra mile to ensure a positive work environment. Given that distinction, employees are more likely to stand behind the company for which they work.

The cost-containment element, the productivity element, and the cultural element are all important. However, companies choose to emphasize different priorities as programs are developed and introduced.

Some of the variables that influence program scope are: the size of the business, the nature of employee's work, the commitment of upper management, and the resources available to manage a program. Generally, health-promotion programs will fall into one of the three categories listed below.

The first is the fragmented, single-intervention program. The second is a categorical activity set, focused on one of the major causes of morbidity and mortality (e.g., cardiovascular disease or cancer). And the third is the comprehensive approach used by such companies as Johnson & Johnson (Wilbur, 1983) and AT&T (Bellingham, Johnson, & McCauley, 1985), which is an organizationally based program grounded in health risk appraisals and including both physical and psychological interventions.

To create a framework for the rest of the discussion in this chapter, the AT&T Total Life Concept (TLC) will be detailed and then this model will be used to illustrate some specific challenges to industry.

In January 1982, AT&T made the historic announcement that it had settled an eight-year antitrust suit with the Department of Justice by agreeing to divest the Bell Operating Companies—which represented about 80 percent of AT&T's total assets.

Top management recognized the enormous impact this would have on the AT&T employee population, and in the spring of 1982, the "Management of Change Committee" was established to address the many divestiture-related human-resource concerns. Out of this effort came the AT&T health-promotion program called "Total Life Concept."

AT&T was fortunate at that time to have a nucleus of leadership that was genuinely concerned about the health and well-being of the employee body as they faced an unprecedented corporate upheaval. But, without a substantial business case for a strategic approach to address employee well-being, they were reluctant to give their approval for a multimillion dollar project. They wanted evidence that there would be some positive return on the corporate investment in the future.

On that note, the Health Affairs Organization compiled a proposal which documented the dramatic increase in the costs associated with the provision of health care to the employee population and further showed how our proposed project would contribute to the goals of containing those costs while at the same time meeting the health and well-being needs of employees during and beyond the divestiture era.

The proposal, endorsed by the AT&T Communications' officers in 1983, overviewed both national and AT&T health-care issues. Prevailing national concerns were as follows:

- Health-care costs had risen 820% per person since 1960.
- Between 1950 and 1982, total health-care expenditures had risen from $12.7 billion to $433 billion per year.

- In 1983, health-care costs were 10.5% of the GNP (Opatz, 1986).

While the AT&T officers were sympathetic to the national health-care issues, their concerns about corporate trends were even stronger. The information on these trends proved to be both dramatic and convincing.

- From 1981–1983, the average rate of increase of insurance health-care premiums was 20% per year.
- Health-care costs paid by the business sector threatened future corporate profits (Herzlinger, 1985a).
- In 1983, 11% of net corporate profits went to health-promotion programs, as compared to 24% of net profits expended on health insurance (Herzlinger, 1986).

However, it was not so much the problem statements but the projected program benefits that won the support of the AT&T executives. It became clear through the listing of potential program benefits that health promotion was one way to invest in the corporation's human resources while at the same time securing a return of the dollars invested. Benefits cited were:

Long Term

Reduced health care costs
Reduced disability
Reduced absence
Reduced premature deaths.

Short Term

Reduced employee health risks
Increased employee satisfaction
Improved employee attitude
Improved quality of life
Increased performance potential
Improved energy and creativity
Improved interpersonal relations.

With approval in hand, the development of Total Life Concept began, grounded in a commitment to a scrupulous design and a systematic implementation process.

TABLE 3.1
A Systematic Approach to Worksite Programming

Planning	Development	Introduction	Implementation	Evaluation
Baseline analysis and objective setting	Training programs and lifestyles components	Orientation and health risk appraisal	Skills and support-based interventions	Process Impact Outcome

The process for the pilot study was a five-step process, culminating in the two-year, 1985 evaluation of the program's effectiveness.

The five steps included planning, development, introduction, implementation, and evaluation. They are laid out in Table 3.1.

The time spent in planning establishes the basis for proceeding with both data and direction specificity to the employer and the employee. The activities of planning include:

• Formulation of mission and goals
• Objective setting
• Hypothesis development
• Analysis of health-care costs
• Assessment of employee population to establish health attitudes, behaviors, skills, and knowledge
• Assessment of corporate culture and norms and their impact on employees' health.

Developmental needs arise from the assessment phase and incorporate tasks around finalizing an evaluation design, selecting marketing strategies, forming leadership and advisory committees, and finally defining the core content and elements of the program components as well as the associated training packages.

Introduction is the phase missing from most programs. Often there is a leap from development into implementation, and there is, in essence, a missed opportunity to impart the mission, direction, and values of the program that is about to unfold. Total Life Concept places great emphasis on the messages of total well-being, planning for wellness, and understanding particular health risks as related to the global picture of "good health."

Implementation should be viewed on a continuum, as it is an aggregate of committee activities, health/lifestyle interventions, organizational interventions, and environmental interventions which occur and recur on a cyclical basis.

The core TLC components are:

Exercise

Back care

Weight management

Smoking cessation

Blood pressure control

Cholesterol/nutrition monitoring

Cancer screening/awareness

Stress management

Interpersonal communications.

Each component is a skills-based lifestyle module with a balanced integration of both substantive information on the particular health topic and support-building and support-seeking strategies. The length of the modules ranges from 4 to 12 weeks, each session lasting approximately one hour.

Evaluation variables selected for TLC were:

Health and job attitudes

Health risks

Health behaviors

Program process (participation)

Cost-benefit analysis.

The AT&T evaluation, carried out by a team of evaluation specialists from General Health, Inc., Project Hope, AT&T Business Research, and Possibilities, Inc., documented statistically significant changes in the areas listed above. The details of these analyses can be found in two articles about the AT&T program (Bellingham, Johnson, McCauley, & Mendes, 1987; Spilman, Goetz, Schultz, Bellingham, & Johnson, 1986).

Since the actual results are not the focus of this chapter, it should suffice to reiterate that positive results were the end product of this process-oriented approach. With that framework as a foundation, it is time to return to the challenges we all face as we move into a new era of health promotion.

In the truest sense, no planning can begin without senior management buy-in. If senior management views health promotion as nothing more than another health program, the complete commitment necessary to implement a comprehensive program is likely to be absent. But, if executive leadership takes the time to become fully aware of the cost-management, productivity, and cultural advantages of health promotion, they will be more likely to commit the company to a long-term, comprehensive process.

In addition, the support of union leadership can play an important role in promoting the programs. Unions that understand the nature of the programs will endorse the activities and encourage members to participate. They may even elect to use their bargaining power to support the process. For example, while negotiating a benefits package, union leadership may relinquish a demand for lower deductibles in exchange for full and equitable access to the health-promotion process.

For the occupational health professional, the challenge is to gain that executive and union commitment, to deliver productive programs in a systematic and cost-effective fashion, and to develop techniques to evaluate those programs and report results in order to ensure continued support.

Another challenge is that we need good scientific data to justify the continuation of health-promotion activities. Substantive, well-executed studies are needed; studies that document bottom-line, quantitative results. This is not to downplay the qualitative benefits of health promotion; but in the business environment where the decision-making process at some point inevitably drives to the bottom line, the translation of effect to dollars is needed to "close the deal."

Today, with the growth of foreign competition and the fear of a pending recession, the trend in industry is to cut costs wherever possible. As a result, occupational health professionals are constantly being asked to justify the cost effectiveness of their programs. In everything we do, we invariably face the question, "Is this a good place for this business to invest its limited dollars— and, if so, why?"

Without adequate data to buttress our response, the task becomes a difficult one, particularly in small- to medium-sized companies where every dime is subject to intense scrutiny.

With that kind of budgetary constraint, many companies cannot justify the expense of wide-ranging scientific studies. For the most part, on-site staff who can devote themselves to this type of work are not available. And even if they were, studies that are confined to a single population are not likely to be generalizable to other work populations.

The first step in rising to the so-called "empirical" challenge is to initiate discussion about the ways in which the scientific community can collaborate with industry to execute studies across organizations and establish a data base that would yield more generalizable evaluations.

In pursuit of that goal, there are two major issues around which some concessions will have to be made. The first issue is program design. Since no two health-promotion programs are exactly alike, some measures of comparability around program activity will have to be established.

The second issue is to recognize that businesses are not laboratories; therefore, it is very difficult to establish and maintain pure control groups in the worksite environment. There are ethical constraints, professional integrity, and needs of the business, all of which need to be factored into program delivery. Many of these factors work against the "pure scientific" protocols that sort out the differences between study and control groups.

The closing message is that in the evaluation of occupational health-promotion programs, rigorous scientific standards can act as the greatest barrier. Recognizing and accepting this as a fact, the peer groups involved in worksite health-promotion evaluation may have to settle for less. "Settle for less" here implies looking at a quasi-experimental design, recognizing the limitations, stating the assumptions, and statistically controlling for variables that cannot be accommodated in the design.

Health promotion needs to build its body of research on cost effectiveness, behavioral strategies, health status changes, productivity improvement, attitudinal changes and more. We occupational health professionals invite the researchers, the scientists, and the academicians to join us in our pursuit of the science of health promotion.

REFERENCES

Bellingham, R., Johnson, D., & McCauley, M. (1985). The AT&T Communications Total Life Concept. *Corporate Commentary, 1,* 1–13.

Bellingham, R., Johnson, D., McCauley, M., & Mendes, T. (1987). Projected cost savings from AT&T Communications' Total Life Concept (TLC) process. *In J. P. Opatz (Ed.), Health promotion evaluation: Measuring the organizational impact* (pp. 35–42). Stevens Point, WI: National Wellness Institute.

Herzlinger, R. E. (1985a). How companies tackle health care costs: Part I. *Harvard Business Review,* July–August, 69–81.

Herzlinger, R. E. (1985b). How companies tackle health care costs: Part II. *Harvard Business Review,* September–October, 108–120.

Herzlinger, R. E. (1986). How companies tackle health care costs: Part III. *Harvard Business Review,* January–February, 70–80.

Kristein, M. M. (1983). How much can business expect to profit from smoking cessation? *Preventive Medicine, 12,* 358–381.

National Survey of Worksite Health Promotion. (1987). Washington, DC: Office of Disease Prevention and Health Promotion, Health Information Center.

Opatz, J. P. (1986). Wellness is a cost containment strategy. *National Safety and Health,* June, 66–70.

Spilman, M. A., Goetz, A., Schultz, J., Bellingham, R., & Johnson, D. (1986). Effects of a corporate health promotion program. *Journal of Occupational Medicine, 28,* 285–289.

U.S. industrial outlook. (1987). Washington, DC: U.S. Government Printing Office, Pub. No. 003–008–00200–5, Superintendent of Documents.

Wilbur, C. S. (1983). The Johnson & Johnson Program. *Preventive Medicine, 12,* 672–681.

4

Work-Site Health Promotion: The European Experience, With Particular Emphasis on the Federal Republic of Germany

Johannes C. Brengelmann
Department of Psychology
Max-Planck-Institute for Psychiatry

Work-site health promotion is a developing concept in the sense that its boundaries are extending to new behavioral areas. Initial attention to the prevention of cardiovascular disorders has expanded to include a variety of physical and psychiatric disturbances. The emphasis is no longer on prevention of illness alone, but also on organizational effectiveness which is known to be impeded by stress. Occupational stress may result from job conditions, role ambiguity, and social interaction. Many behavioral problems at work stem from individual predispositions to stress. Recently, research has also concentrated on the spillover of strain produced in the family, and on the question of social support.

In Europe, at least, research on health promotion is not of a high standard. Researchers are more aware of research activities carried out in the U. S. than in their neighboring countries. While European research centers are willing to produce behavioral research in cooperation with work settings, there are few research programs carried out systematically within the work site.

These are the main reasons why the factual part of this chapter concentrates on West Germany, where some interesting developments have taken place. The focus is limited to behavioral programs in or close to the work site and to the exclusion of poorly organized community programs, activities of self-help groups, or research projects based outside the work site. This limitation reduces the chapter to a report on a few controlled studies instead of undertaking the unrewarding task of reviewing the many abortive or uncontrolled programs.

Traditionally, ''work medicine'' (Arbeitsmedizin) has focused on health aspects at work or occupational health in most of Europe, but these efforts were not conceived of as behavioral and were not conducted by behaviorally trained personnel. In recent years, the term "work climate" has received research attention. Sweden, Great Britain, the Benelux, and Germany are among the countries where such work has been produced. The work-site diagnosis of psychosocial stress, occasional availability of stress-management training, university-based behavioral training to help prevent or control high blood pressure or the typically German tradition of visiting a health spa are varied examples of diverse activities to combat stress. Noteworthy is the recent establishment of six behavior-medical-rehabilitation hospitals in Germany. This development is promising, but overall, behavioral medicine efforts lack cohesion and the acceptance of a unifying theory. Many different schools of thought help to muddle the picture, offering assistance without concern for a scientific preparation of health-promotion programs based in the work site.

Our hope for the future rests on the development of behavior therapy associations which began to establish themselves in Europe 20 years ago and, moreso, on institutions of behavioral medicine which are now new on the scene. Behavior therapy has initially been occupied with psychiatric, educational, and medical problems, in that order. The door to occupational settings and public institutions is being opened reluctantly. This may be due to the fact that expertise, power, and influence in industry are distributed differently from the traditional areas of health, education, and welfare.

In West Germany the origins of the present health programs in the work site can be traced to the following sources:

- the development of behavior therapy and behavioral medicine under the leadership of the Max-Planck-Institute of Psychiatry in Munich since 1966;
- the founding of several national and local behavior therapy organizations during the last 20 years;
- the operation of a nonprofit, government-supported private Institute for Therapy Research and Training; and
- the impressive spread of behavior therapy at the university level during the past decade.

With regard to health promotion in the work site, these institutions have evolved intimate interactions with occupational and family settings, thereby paving the way for change.

The following four programs will be discussed because they are comparatively well controlled.

- a public health center program for the prevention of cardivascular disease (CVD);

```
┌─────────────────────────────────────────────────┐
│   PH CENTER PROJECT: CVD PREVENTION               │
│                        in the community           │
│                                                   │
│              I. Experimental Model   II. Transfer Model
│                                                   │
│  Location:   Mettmann Cty           9 counties (35 centers)
│              (5 centers)                          │
│  Duration:   Dec 1977 — Sept 1982   Oct 1982 — June 1984
│                                                   │
│  Goals:      Establish              Test program  │
│              programs against       effectiveness in 2 counties
│                — smoking                          │
│                — overweight         Describe results of transfer
│                — lack of exercise                 │
│                — stress             Analyze favorable and
│                                     unfavorable conditions
│                                                   │
│              Test effectiveness of  Assess and solve problems
│              programs               of transfer  │
└─────────────────────────────────────────────────┘
```

FIG. 4.1. A public health center project for the prevention of cardio-vascular diseases.

- a general health center in a large industrial organization;
- the use of policemen as mediators in stress management and communication training; and
- the installment of a behavioral employee counseling program.

The public health center program for CVD prevention, shown in Figure 4.1, was open to anyone and was placed in the community surrounding the work sites because health experts felt that this location was preferable to the work site itself. The program situated at Mettman County, just outside of Duesseldorf, was supported by a local public insurance company and the Federal Ministry of Labor and Social Affairs. Five walk-in centers were set up between 1977 and 1982 which offered free advice and behavioral training on smoking, overweight, lack of exercise, and stress reactions. The program was well accepted by the clients and the news media, as well as by the local medical practicioners. The services were subsequently extended to nine other German counties with 35 centers in all. Here, research concentrated on problems of dissemination in view of the expectation that the model might be extended to all of the country. The major drawback of this well-conducted and documented research (Wengle, 1985) is that while the behavioral change in the four program components was tested, no funds were forthcoming to assess the program effects on health in general. This problem has continued to exist.

Figure 4.2 indicates that the first industrial health center was established in 1986 to serve the needs of the employees of Bayer-Leverkusen, a large pharamaceutical and chemical firm. This example is cited for three reasons, the first being historical relevance. To my knowledge, this is the first industrial prevention center in Germany planned according to behavioral principles. Second, the center offers services for a great variety of problems of both employees and their families. Third, and regretfully, no provision was made for the assessment of efficacy and maintenance of procedures, so that no further evidence can be presented. This may serve as an example of how *not* to proceed.

The development of stress-management trainers (mediators) in the police force is an example of a well-controlled experiment. The aim was to enable as many police officers of Northrhine-Westphalia as possible to handle their daily duties more effectively (i.e., with less stress reaction and improved communication). Figure 4.3 reveals how this was achieved in less than four years. In December 1983, stress-management training procedures were demonstrated to police officials. As a result, eight police officers were selected to undergo a 17-day training in 1984 over a period of four months. They became the first mediators of behavioral training in the police force and themselves trained 92 colleagues in a 22-day training course in 1985.

Finally, a special task group of about 90 police trainers was established in order to provide stress management and communication training for 2,500 police officers each year. The aim is to extend the service of a three-week training to all officers of the force who request it. The training takes place in the refurbished Schellenberg Castle at Essen which has 240 beds and 18 training rooms. The long-term goal is to revise training and assessment procedures on the basis of their demonstrated efficacy, as well as to add leadership training. The program has been well received by the media. Police representatives claim a drastic reduction in the use of guns and in the number of complaints by

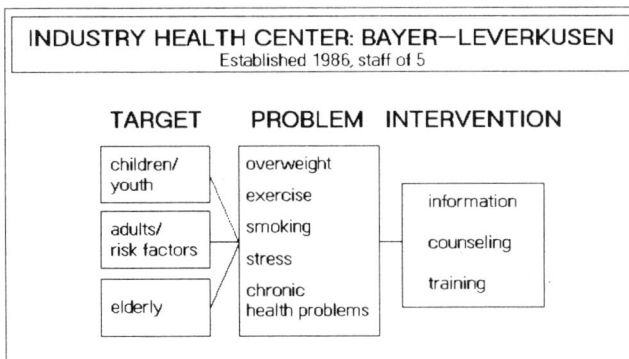

FIG. 4.2.   Establishment of the first behavioral health center in industry in 1986.

```
┌─────────────────────────────────────────────────────┐
│  ┌──────────────────────────────────────────────┐   │
│  │  POLICE FORCE:   WORK─SITE PROBLEMS  │           │
│  └──────────────────────────────────────────────┘   │
│                                                       │
│  Goal:      More effective handling of work─site problems
│             (fights, demonstrations, riots as well as normal
│             duty operations)
│
│  Step 1:    — Demonstrate SMT (1983)
│  (1983)
│  Step 2:    — Identify and categorize stressors
│  (1984)     — Train officers to be police trainers (mediators)
│             — Establish effectiveness criteria
│
│  Step 3:    — Mediators train other police officers without
│  (1985)       supervision
│             — Check training effectiveness
│             — Establish training center (240 beds, 18
│               training rooms)
│
│  Step 4:    — Routine training of 2,500 officers per year
│  (1986)     — Revise stressors, training, criteria
│
│  Step 5:    — Check effects of training in problem situations
│  (1987)     — Check long─term effect of training on personal
│               effectiveness
│             — Plan training for inclusion in general police
│               education
│             — Plan revision of police image and ministerial
│               policy
└─────────────────────────────────────────────────────┘
```

FIG. 4.3. Stress-management training in the police force by means of mediators.

citizens at the police stations. These claims are being investigated. Figure 4.3 also states some longterm goals of this program. Authorities must learn that behavioral programs require continued maintenance, that behavioral principles are to be used at all levels of police education, and that ministerial policy can gain considerably by incorporating behavioral theory and practices.

The immediate and long-term effectiveness of the stress-management training just described has been determined by means of subjective and objective assessment techniques. The results are being published (Brengelmann, 1987; Brengelmann Bruns, 1990). One particular finding is shown in Figure 4.4, which reflects Step 2 of the program (compare Fig. 4.3) in which the first eight mediators trained 92 fellow officers to become stress-management trainers. The success of this is assessed by means of a Stress/Coping questionnaire which, in its final version, consists of 300 items yielding 50 primary and 10 secondary factors relating to personal effectiveness (success orientation) and ineffectiveness (stress, incompetence), as shown in Figure 4.5. Results reveal that self-competence is not significantly altered in the course of this training. Social competence is improved, and incompetence is reduced. All stress reactions, including negative evaluations, are greatly reduced by training and, apparently, normalized. Similar results are obtained by means of independent measures of life quality, as well as of Type A which are not reported here. The two most significant conclusions are that mediators can be competent behavioral change agents and that training changes specific behaviors instead of behavior across the board. This should help in paving the way for the increased use of mediators, which is certainly needed.

The last program to be discussed is shown in Figure 4.6. It concerns behavioral counselling of employees in the Ministry of Northrhine-Westphalia. Industrial and public organizations alike have become alarmed by problems associated with alcohol abuse and, subsequently, with other behavioral problems. Employee health-assistance programs in German industry have been in existence for over 100 years. They have been conducted by internal personnel, medical doctors, social workers, and/or nurses. The present program is novel in that behavioral psychologists train regular employees to become social counselors (in German SAP = sozialer Ansprechpartner). The program is de-

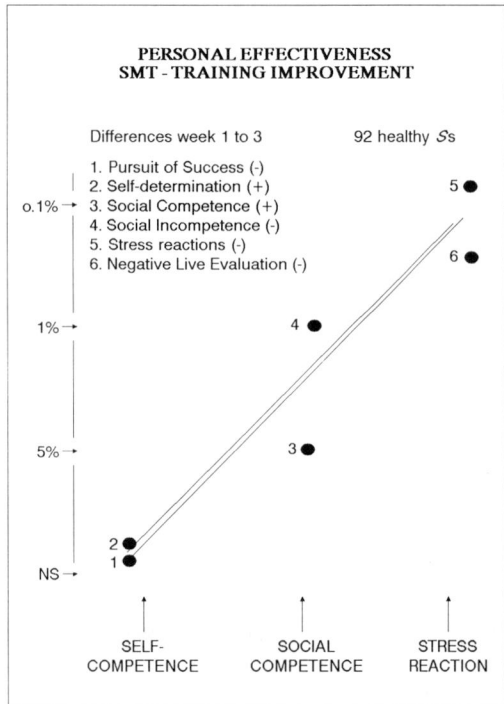

FIG. 4.4. Change in personal effectiveness due to stress-management training.

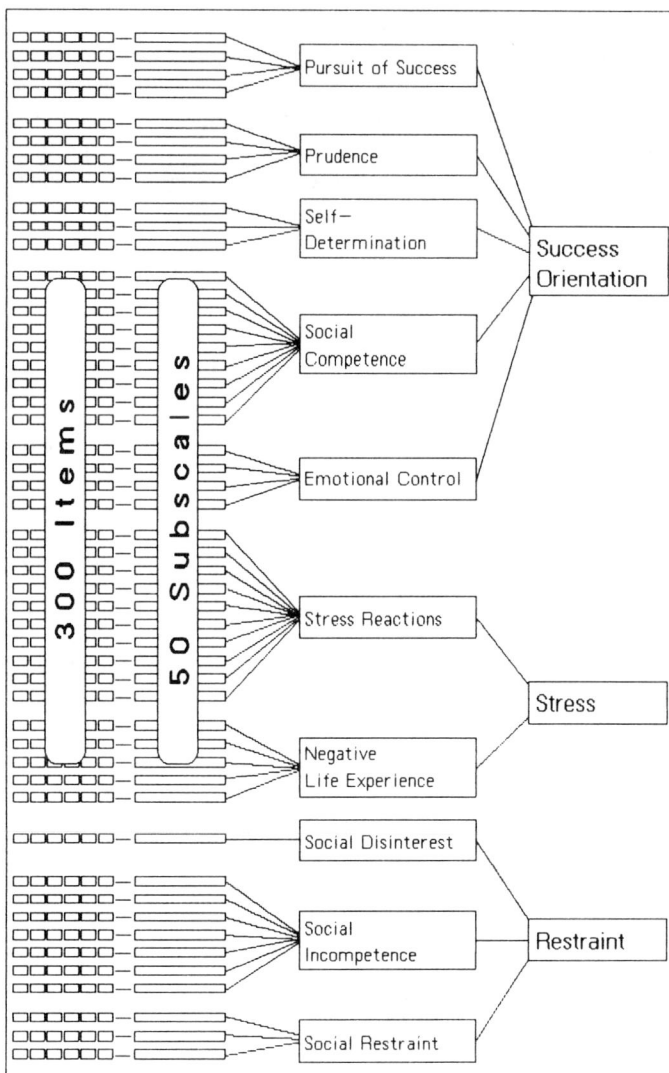

FIG. 4.5. SCOPE: A questionnaire concerning success orientation, stress, and restraint.

signed to proceed in three phases. First, the variety and prevalence of individual and organizational problems are determined. Particular emphasis is placed on the flow of rewarding, neutral, and punishing influences within the organization. In one study it was found that stressors impinging on cancer therapists stem from their colleagues as often as from the patients (Ullrich, 1987). In an analysis of a behavioral medical hospital currently in progress, both

facilitatory as well as stressing organizational influences are being investigated. In the second step of our counseling program trainees learn to
recognize their work-site problems and to act against them preventively. Training was provided by experienced psychologists during 1987 and 1988 in 10
blocks of 3 days each, as is described in Figure 4.6. The main tasks of the
third step are to provide several one-day refresher courses until deemed unnecessary, to fade out supervision, and, most importantly, to revise procedures
and philosophies on the basis of the experience gained.

To conduct complex programs at work is quite an exciting experience.
However, the most rewarding aspects have been:

refinement of measurement of subjective variables,

emphasis on positive behaviors, in contrast to stress reactions, and

relating personal and organizational efficacy to behavioral life philosophies.

These points will be dealt with in the remainder of this paper. The development of the SCOPE questionnaire (already shown in Fig. 4.5) has provided
insights regarding the success/stress balance. This differentiates reliably between a number of medical, psychiatric, and healthy groups; for example, between different types of managers. Heart rate varies significantly with success
orientation, but not with stress reactions. Such results have encouraged us to

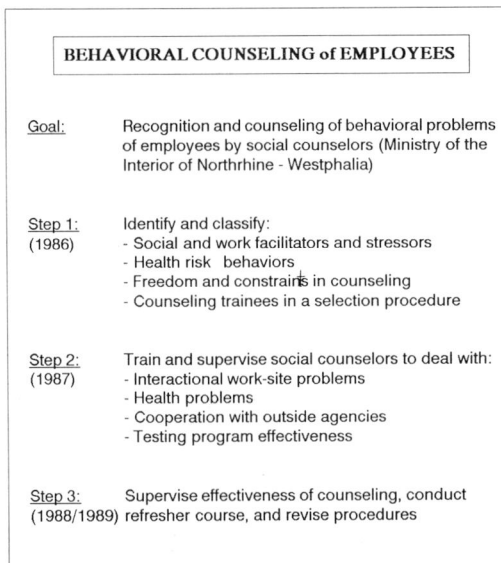

BEHAVIORAL COUNSELING of EMPLOYEES

Goal: Recognition and counseling of behavioral problems
of employees by social counselors (Ministry of the
Interior of Northrhine - Westphalia)

Step 1:   Identify and classify:
(1986)   - Social and work facilitators and stressors
- Health risk   behaviors
- Freedom and constrains in counseling
- Counseling trainees in a selection procedure

Step 2:   Train and supervise social counselors to deal with:
(1987)   - Interactional work-site problems
- Health problems
- Cooperation with outside agencies
- Testing program effectiveness

Step 3:   Supervise effectiveness of counseling, conduct
(1988/1989) refresher course, and revise procedures

FIG. 4.6. Behavioral counseling by means of employees trained as social counselors.

LIFE—QUALITY OUTCOME CRITERIA
Cross—validated factor analysis

SUCCESS   STRESS

personal effectiveness

initiative   stress reactions
positive living   pessimism
optimism   Type A
internal control   external control
life success   health risks

family effectiveness

freedom of expression   conflict
freedom of action   restlessness
mutual reinforcement   worries
conflict solution   overload
discipline   material striving

work—site effectiveness

openness   time pressure
identification   work shift
information   conflicts
self—determination   output overdemand
remuneration   ability overdemand

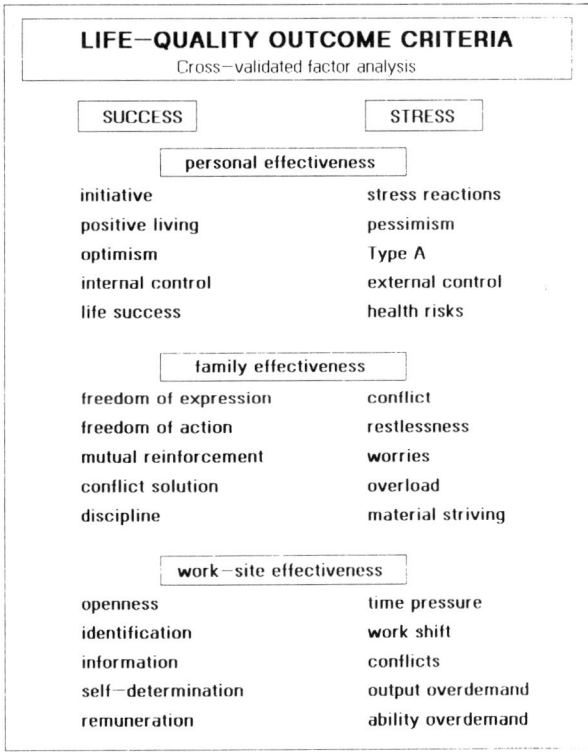

Fig. 4.7. Life-quality questionnaires for success and stress experienced on the personal, family, and work-site levels.

develop further life-quality parameters for success and stress on the personal, family, and work-site levels, as is shown in Figure 4.7. The factors listed are of second order, while ''success'' and ''stress'' are of third order. The factorial method used is the principal component analysis with varimax-rotation.

One advantage of this procedure may be seen from Figure 4.8, where personal effectiveness is compared with group effectiveness, or social climate, in the family and in the work-site ($N > 200$ per comparison). Personal success correlates poorly with social effectiveness in the family or in the work-site, while personal stress correlates very highly with social stress. It is concluded that successful behaviors in the sense of effective self-competence and social competence are specific to situations, while stress reactions are rather unspecific. It follows that stress reactions are ineffective variables, on account of their unspecificity. It is, furthermore, interesting to note that personal and social effectiveness interact significantly with stress in the family, but not in the work-site. This may be explained by the tighter social interaction in the family.

PERSONAL EFFECTIVENESS VERSUS GROUP CLIMATE

PERSONAL EFF. (SCOPE)	CLIMATE	
	SOCIAL	STRESS
FAMILY		
SUCCESS	10	- 35***
STRESS	- 43***	68***
WORK-SITE		
SUCCESS	17*	01
STRESS	- 12	51***

Significance levels: *5%, ***0.1%

FIG. 4.8. The relationship of personal effectiveness to social effectiveness in the family and at work.

DIFFERENTIAL EFFECTS OF ANXIETY MANAGEMENT TRAINING AND POSITIVE BEHAVIOR TRAINING

	ANXIETY MANAGEMENT TRAINING			POSITIVE BEHAVIOR TRAINING		
INTERNAL STRESS	post	1 mo.	4 mo.	post	1 mo.	4 mo.
Sleep/physical problems (-)	*	**	*	*	***	***
Anxieties (aging, dying, health) (-)	*	**	***	*	**	**
Self-criticism (-)	NS	*	*	*	***	***
Depression (-)	NS	*	*	NS	***	***
MANIFEST BEHAVIOR						
Social incompetence (-)	NS	NS	*	**	***	***
Assertiveness (+)	NS	NS	NS	***	***	**
Relaxation (+)	NS	NS	NS	**	***	**
Social pleasure (+)	**	NS	**	***	***	NS
Work strain (-)	*	*	NS	*	***	***

Significance levels: NS = not significant, *5%, **1%, ***0.1%; AMT and PBT each 23 *S*s.

FIG. 4.9. Positive behavior training is more effective than anxiety management training.

46

One further example for the importance of positive behavior, as compared to stress or anxiety, is provided in Figure 4.9. Medical patients suffering from chronic stress disorders, mostly of the cardiovascular or gastrointestinal type, were randomly assigned to either anxiety management training (AMT) or positive behavior training (PBT). AMT concentrated on the reduction of negative psychological symptomatology, while PBT ignored discussion of personal problems and relied only on the training of positive behaviors. Duration, intensity, and format of training were otherwise comparable. It is evident that AMT is fairly effective with regard to the reduction of stress symptoms, but not so much with regard to positive behaviors. By contrast, PBT is more effective in all but one respect (anxieties). The conclusion is obvious—the enhancement of positive behavior is the method of choice.

The points made in relation to the last three figures are central to dealing with work-site problems; at least as we use them in our programs.

But, in order to grasp the wider philosophical context of behavior at work, it is advantageous to know the goals and means of self-realization or the philosophies of life in the population at large. This is shown in Figure 4.10. On the basis of a representative poll taken in West Germany by the Social Democratic Party ($N = 5,000$), a scale for attitudes of life was constructed and applied to over 600 persons in various walks of life. Factor analysis yielded what may be called four philosophies of life. The first factor to emerge was termed ''Progress.'' It represents the quest for personal advancement and prestige proposed in reasonable, objective terms. The possession and enjoyment of material things (ie., the pleasures of life) appear to be important. ''Security''

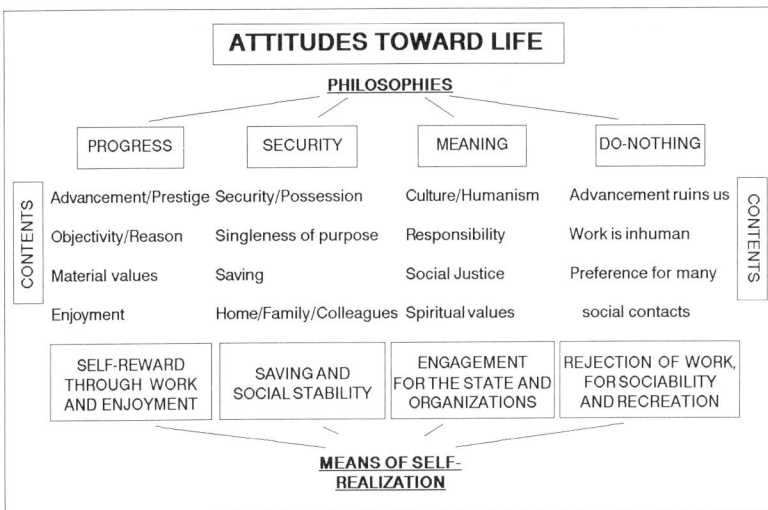

FIG. 4.10.   The structure of attitudes toward life.

stands for hard work and for the purpose of securing the necessities of life: a home for the family, investments, and sharing life with a few reliable friends or colleagues. "Meaning" characterizes people striving for "higher" values of the humanistic or spiritual kind. They surround themselves with the air of responsibility and social justice. The fourth factor was tentatively termed "Do-nothing," but also contains strong elements of furstration at having to live in a society with the wrong values. Subscribers to this philosophy shun work or consider work to be evil and inhuman. Their sense of living consists of rejecting bourgeois life and spending the day leisurely without any commitment. These four philosophies are achieved via different reward systems, as follows:

Progress   = self-reward through advancement and enjoyment with no strings attached

Security   = small-group recognition for material possession obtained by excessive work

Meaning   = large-group recognition and dedication to idealistic goals

Do-nothing = rejection of societal rewards, enjoyment of momentary need fulfillment.

This system of behavioral philosophy opens doors for discussion for which there is no time or space in the present context. However, it is known that progress correlates strongly with success orientation (SCOPE) and negatively with stress. It may be termed the prototype of positive self-reinforcement. Security correlates significantly with success, as well as stress. It relates to Type A. Meaning correlates negatively with success, while Do-nothing relates to stress only. These and other analyses place the behavioral observations in the work-site into a wider framework of understanding. Maybe that is a unique or typical European contribution to the problem under consideration, and it enriches programs for the enhancement of work-site behavior.

REFERENCES

Brengelmann, J. C. (1987). *Stressbewältigungstraining 1*. Frankfurt, West Germany: Peter Lang.

Brengelmann, J. C., & Bruns, G. (1990). *Stressbewältigungstraining 2*. Frankfurt, West Germany: Peter Lang.

Ullrich, A. (1987). *Krebsstation: Belastungen der Helfer*. Frankfurt, West Germany: Peter Lang.

Wengle, E. (Ed.). (1985). *Modellversuch Gesundheitsberatungsstellen bei der AOK für den Kreis Mettmann (AOK Projekt)*. Abschlußbericht. Bonn, West Germany. Forschungsbericht Nr. 124, Band 1, Gesundheitsforschung BMA (Federal Ministry of Labor).

5 Control Data's StayWell® Program: A Health Cost Management Strategy

William S. Jose II
OverView Consulting

David R. Anderson
StayWell Health Management Systems, Inc.

Control Data's StayWell program is a comprehensive health-promotion program designed to manage health-care costs and improve productivity by reducing the level of lifestyle risk among employees and their families. The program was initially developed for Control Data's employees. Today the StayWell program is also available [from StayWell Health Management Systems, Inc.] to other organizations through a nationwide network of over 50 authorized distributors.

Control Data began development of the StayWell health-promotion program in 1978. At that time, a major commitment was made to evaluation of the program. The motivation for this major investment was twofold. Initially, there was a determination to provide an effective program at Control Data that would enhance employees' health and productivity, and at the same time reduce health-care costs. Later, when external marketing of the StayWell program began, there was a well-recognized need to be able to demonstrate the value of the program to prospective corporate customers. This chapter provides an overview of the StayWell evaluation to date (1988) and highlights some of its more important findings.

STAYWELL PROGRAM DESCRIPTION

Several components of the StayWell program were piloted in 1979. Phased implementation to Control Data work sites began in 1980. By the end of

---

StayWell is a registered trademark of StayWell Health Management Systems, Inc.

1985, the StayWell program had been implemented in most major U.S. Control Data facilities.

The StayWell program has undergone tremendous evolution since its initial implementation. Much of this evolution was the direct result of evaluation of pilot materials and processes that identified program strengths and weaknesses. The relatively "mature" program of today (Jose, Williams, & Keller, 1986) is organized around the four key phases of the health-promotion process:

- awareness;
- assessment;
- behavior change; and
- maintenance.

The awareness phase consists of program introduction and promotion. Distinct orientation programs are provided for managers and employees. The orientation for managers is designed to gain support for the program, establish communication and reporting channels, and plan and schedule program implementation. The orientation for employees is designed to inform them of program opportunities and build motivation for program participation. Employee orientation is most effective in group sessions, but may also be accomplished very effectively by direct mail. A videotape presentation provides an additional option for either group or individual orientation.

The assessment phase consists of assessment of individuals *and* the organization. Organizational assessment is accomplished by a sample survey of the organization using the Employee Health Survey, (Jose, Anderson, & Peterson, 1985). This instrument provides information on health risks and health-related attitudes, knowledge, and skills. Because of the sampling procedure, the results are representative of the entire organization. The Employee Health Survey is first administered before the program is introduced to employees. It functions to assess needs, plan implementation strategies, and establish a baseline for ongoing evaluation.

Individual assessment is accomplished using the HealthPath® health-risk profile which consists of a questionnaire and several physiological measurements, such as cholesterol, heart rate, blood pressure, and body composition measurement, as well as other options. The individual HealthPath report ranks the individual in 11 health habit areas, focusing on the three habits most in need of change to reduce overall health risk. It also compares current rankings with those of the individual's previous HealthPath results. An integral part of the individual assessment is an interpretation, which helps to explain the results and motivate the participant to make positive changes in health habits. The interpretation can be accomplished in groups or individually, with or without the intervention of a health educator.

The behavior-change phase consists of a variety of educational programs and services designed to assist the individual in making the first steps toward incorporating healthy habits into his/her lifestyle. The StayWell program currently includes five types of educational components: instructor-led courses in one-hour segments; instructor-led courses in half-hour segments; self-study courses; campaigns; and behavior-change booklets. Educational components are currently available in the following areas:

- fitness
- smoking cessation
- stress management
- driving safety
- nutrition
- weight control
- back care
- health-care consumerism

The maintenance phase consists of activities and services designed to assist individuals in making new habits a permanent part of their lifestyles. It helps to create an environment in the work place that supports and reinforces healthy lifestyles. The *WellTimes* quarterly newsletter is an integral element in keeping the visibility of health-related issues before employees. Action teams are employee-led groups that focus either on activities directly relevant to maintaining positive health habits, or on making the work environment itself a healthier place (e.g., establishing a walking route or providing nutrition labeling in the cafeteria).

The flexibility that the StayWell program provides with its many components and alternative delivery modes helps to meet a wide variety of specific needs. Proprietary analysis and planning tools such as the Employee Health Survey, HealthPath Management Summary, Course Evaluation System, and StayWell Cost Model help decision makers to target organizational needs and measure progress toward program objectives.

EVALUATION DESIGN AND ANALYSIS

StayWell Process Model

The design, development, delivery, and evaluation of the StayWell program are guided by the StayWell process model (Fig.5.1). This model assumes that initial attitude, knowledge, and behavior change occur together, due to the effect of the program. It makes no assumptions about how they are causally related. To the extent that initial changes are evaluated positively by the individual and supported by the environment, the changes will be maintained and incorporated into the individual's lifestyle on a more long-term basis. The model acknowledges the inherent long-term nature of the behavior-change

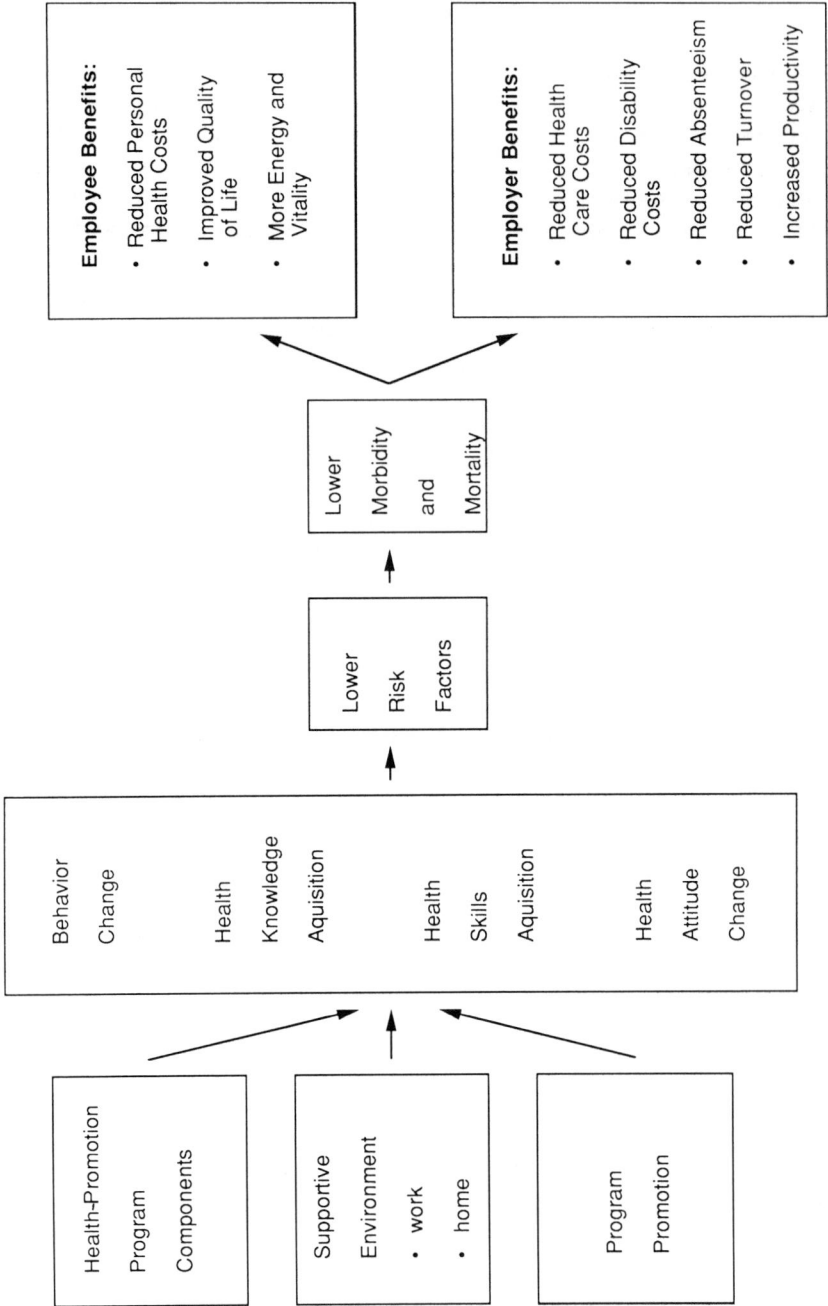

FIG. 5.1. StayWell Process Model.

process, and emphasizes the importance of environmental components, including both social-support and physical work-site alterations.

The StayWell process model is comprehensive in its treatment of the work-site behavior-change process, in that it encompasses individual knowledge, skills and attitudes; incorporates a behavior-change process; and emphasizes both internal and external factors in establishing long-term behavioral changes. In addition, it recognizes the link between behavior change and reductions in risk factors that lower morbidity and mortality, thus reducing employer costs.

Three of the most significant implications of the StayWell model for program design, implementation, and evaluation are that it focuses on:

- behavior change rather than knowledge and attitude change;
- social support as an essential element in successful long-term behavior change; and
- the critical importance of both physical and sociocultural aspects of the work-site environment.

Evaluation Objectives

The StayWell program evaluation objectives are consistent with the process model. Objectives fall into four broad categories:

1. To assess risk status in the eligible employee population initially and in subsequent time periods.
2. To evaluate StayWell components and the overall program to support further development and refinement, including:
   - participation rates and patterns;
   - participant reactions to program activities; and
   - changes in participant knowledge, attitudes, skills, and behaviors.
3. To assess the effect of the StayWell program on long-term risk reduction.
4. To determine the impact of the StayWell program on health-care costs and productivity (e.g., absenteeism).

Assessment of program impact on excess costs is complicated by the time required for behavior changes to become manifest in reduced absenteeism or health-care costs. The impact of wearing seat belts on costs is immediate. However, the cost savings associated with smoking cessation can be expected to take several years to be fully realized (Koop, 1983). Time lags between behavior or physiological change and reduced health-care costs are less well understood in the areas of weight reduction, increased cardiovascular fitness, and lowered cholesterol.

Design Considerations

The ideal approach to evaluating the effect of any treatment or intervention is a true experimental design that permits cause-effect relationships to be clearly established. However, it has not been possible to use true experimental design to evaluate the StayWell program because individual participation is voluntary, entry of individuals into the program is not controlled, and implementation sites have not been randomly selected.

Consequently, the strategy for evaluating the StayWell program has been to use a variety of quasi-experimental designs (Cook & Campbell, 1979). Specific designs have included matched sites, pre-post, time-series, and multiple time-series. The rationale underlying this strategy is that, while each of these designs has certain inherent limitations, the specific weaknesses vary across designs. Thus, to the extent that similar conclusions are reached using several different designs and involving a variety of data, fairly strong conclusions about overall program impact can be drawn.

Data Collection

Data for analysis of the StayWell program come from seven primary sources:

- participant feedback about program components;
- Employee Health Survey;
- health risk profile (i.e., HealthPath);
- personnel records;
- StayWell participation records;
- health-care claim records; and
- paid sick-leave records.

The Employee Health Survey (Jose, Anderson, & Peterson, 1985) has been administered annually since 1980 (except 1984 and 1985) to representative samples of Control Data employees. Employee Health Survey data are not longitudinal since the composition of samples changes from one administration of the survey to the next. Nevertheless, survey results provide a periodic profile of the health of the corporation against which changes may be measured and related to specific StayWell interventions.

Longitudinal data from individual health-risk profiles, personnel records, StayWell participation records, health-care claims records, and paid sick-leave records are stored in a data base and can be linked through employee identification number (Haight, 1985). Identification numbers have been encrypted to protect individual confidentiality. This data base provides a powerful tool for evaluating the StayWell program that, to our knowledge, cannot be matched by any other work-site health-promotion program. The StayWell data base enables evaluators to:

- link risk factors and program participation to health-care claims experience and sick-leave use;
- track individual risk-factor change over time; and
- link risk-factor change to demographic characteristics, StayWell participation, and sick-leave and health-care claim trends.

OVERVIEW OF EVALUATION RESULTS

Participation

The StayWell program has been well received by Control Data employees. As of the end of 1986, 71% of eligible full-time employees were enrolled in the program, and 53% had completed one or more health-risk assessments. In addition, 44% of eligible employees had participated in one or more courses or action programs. Overall, 63% of eligible Control Data employees had participated in at least one StayWell activity.

Enrollment in the StayWell program exceeds 50% for all age, gender, and job categories, except for those 60 years of age or older. The percentage of those enrolled who participate in some component of the program in addition to the health-risk assessment is also over 50% for all age, gender, and job categories. These data indicate that the program has been successful in reaching all age, gender, and job groups. The current success of the StayWell program in attracting widespread participation is due to actions taken on the basis of early evaluation results to strengthen the behavior-change component of the courses, to offer courses in a variety of formats, and to give employees greater ownership in the program.

Participation and program acceptance across age, gender, and job groups is an essential first step in a successful program. An important second step is to assure that at-risk employees are participating in activities relevant to reducing those risks. Particularly satisfying to Control Data is the percentage of high-risk employees who have participated in StayWell activities relevant to reducing their risks. Participation by high-risk employees in a program component relevant to reducing that risk has reached 59% for overweight, 59% for hypertension, 50% for cholesterol, 38% for fitness, and 26% for smoking. The lower participation by smokers may indicate the existence of a resistant high-risk group due to reaction to strong societal anti-smoking pressures. All things considered, the StayWell program is making excellent progress in reaching the high-risk segment of the employee population. Much of this progress came after 1983, when specific actions were taken to better target high-risk participants.

Employee Perceptions

Positive changes in employee perception are important benefits of effective work-site health-promotion programs. Employee perceptions of the StayWell program have consistently been very favorable. In the Control Data Employee Health Survey of 1986, 84% of respondents at StayWell sites indicated that the program helped to improve employee health and productivity; 72% perceived the StayWell program as a valuable benefit; and 68% perceived the program as a cost-saving measure. Over the years, overall satisfaction with the StayWell program among employees has hovered around 80%. Coupled with enrollment of 71%, these ratings clearly indicate that the program is liked and valued by employees and understood as a program that saves money for the company.

Work-Site Culture

Each StayWell program component has been designed so that it can be used effectively on a stand-alone basis. This "unbundled" approach to design is essential to meet the specific needs and resources of the many different Control Data work sites and StayWell participants. Existing data, however, suggest that the program components work together to produce a greater effect than would be expected based on the performance of individual components (Fielding, 1984). This synergy is consistent with the StayWell process model, since the implementation of an ongoing comprehensive program alters the work-site culture relative to health-related knowledge, beliefs, and practices. The altered work-site culture then supports norms directed toward healthy risk-reducing activities, increasing the probability of successful long-term behavior change.

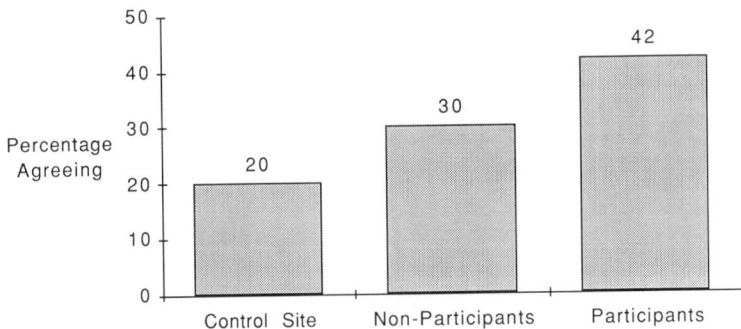

FIG. 5.2. Percentage agreeing that co-workers are improving their health habits. Source: CDC Employee Health Survey; 1983.

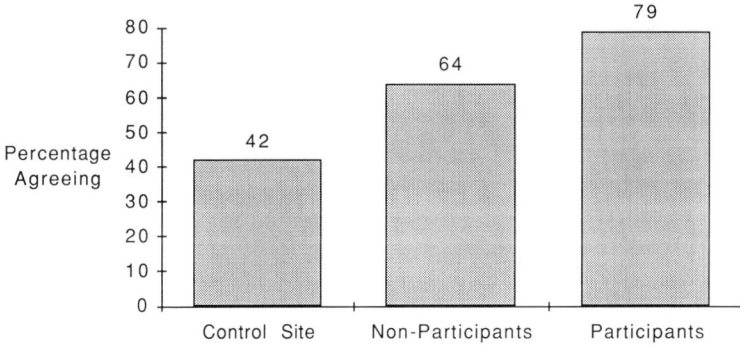

FIG. 5.3. Percentage agreeing that there is support at work for good health practices. Source: CDC Employee Health Survey; 1983.

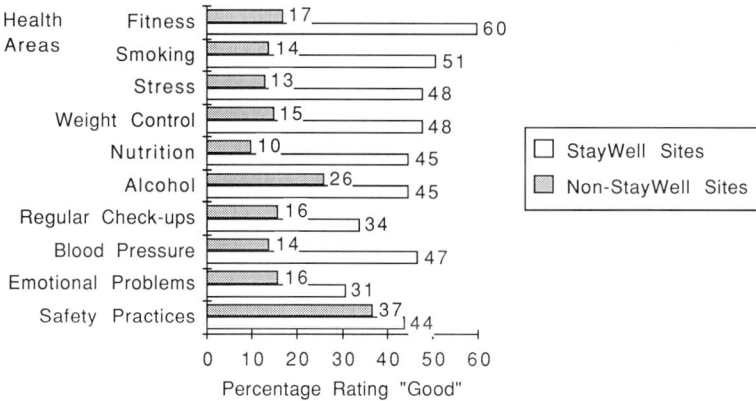

FIG 5.4. Percentage who believe that CDC is doing a good job of supporting employee health practices. Source: CDC Employee Attitude Survey; 1981.

The StayWell evaluation has found, for example, that individuals not participating in the StayWell program, but working at a site where the program is active, are substantially more likely than comparison site employees to agree with the statement, "Co-workers are improving their health habits" (Fig. 5.2). Such a perception is an important indicator of a changing culture relative to good health practices.

Similar evidence of a changing culture relative to health practices emerged when employees were asked to rate support at work for good health practices (Fig. 5.3). Here, again, even nonparticipants at program sites rated support for good health practices higher than did control site employees.

57

Figure 5.4 reports perceptions of corporate support for employee health in specific areas. In all ten areas assessed, many more employees at StayWell sites than at comparison sites thought Control Data was doing a good job of supporting their efforts. Again, these changed perceptions of employees at StayWell sites occurred irrespective of direct participation in the program.

Behavior Change

Early in the program, it was discovered that the original lifestyle change courses relied too heavily on mere dissemination of information. Analysis revealed that participants wanted and needed a more "how-to" orientation. In the absence of such a skills-oriented approach, the desired behavior and lifestyle changes were not occurring. In addition, completion rates were quite low for the initial set of courses.

In 1983, redesigned courses were piloted. Striking results in the desired direction were noted. The overall completion rate jumped from about 50% for the original courses to over 80% with the newly designed courses. Behavior change and satisfaction with the courses also increased. The analyses of course results carried out systematically over the years, and the course modifications that resulted, created a high level of confidence that the courses work; that is, they are effective in helping participants to reduce risk levels.

Effective courses are an important component of successful work-site health promotion, but they can only be successful if employees participate in them. Analysis of participation patterns indicated that certain job groups had significantly lower participation in lifestyle change courses. These job groups included sales, marketing, and customer-service representatives, whose frequent travel made it difficult for them to maintain regular participation in a multi-session course.

To meet the needs of these individuals, self-study courses were developed. The self-study courses, while missing the potentially motivating interaction of instructor-led courses, provide all the materials needed to assist a person in making positive lifestyle and behavior changes. As such, they are also well suited for use at small or remote sites where it is difficult or expensive to provide instructors. In addition, many individuals who do not like to attend courses or who feel awkward in group situations find the self-study courses useful.

Evaluation of the self-study courses indicates that they are well liked by participants and effective in initiating behavior change. The overall completion rate for the five self-study courses is about 30%. While incentive programs have been used to produce much higher completion rates, even the 30% overall rate compares favorably with similar minimum-contact interventions which typically have a high degree of non-completion. Eighty-seven percent of the completers rated the self-study courses "good" or "excellent." Additionally, short-term compliance among completers was very good. For example, completers of the weight-control course lost an average of 11.4 pounds, 77% lost weight, and 61% exercised more often. Thirty-four percent of those completing the smoking course quit smoking, and an additional 43% reduced their smoking.

Evaluation of the implementation process also indicated that the one-hour courses, typically held during lunchtime, were not meeting the needs of production workers whose lunch periods were shorter. Therefore, courses were redesigned to meet this need. A series of instructor-led courses based on half-hour presentations were made available. These have proven to be effective in meeting the needs of this target population.

In a further response to the identified need for emphasis on long-term behavior change, action teams were added to the StayWell program to enhance and strengthen the social-support and environmental components identified in the process model. The social support provided helps individuals to maintain behavior changes initiated in the lifestyles change courses. This support is often essential in the early maintenance phase of behavior change. Analysis of individuals' reactions to action teams indicated that the most effective teams were not led by course instructors, but by employees. Leadership by employees created more of a sense of ownership and responsibility, and, hence, more effective behavior change and maintenance. As a result, employee involvement in planning and implementing action teams is sought to ensure that these activities meet employee needs. Because the support and reinforcement of positive behavior change is so important during the critical transition period between course completion and internally motivated long-term lifestyle change, action teams are considered a key element in the success of the Stay-Well program. This support has helped to reduce the widely reported recidivism of other programs with similar goals.

Organizational Risk Reduction

Numerous changes in Control Data have had an effect on lifestyle-related health risks. Three major impacts have been the StayWell program, changes in the benefit plan, and certain corporate policy changes. Components of the StayWell program have been described in some detail above. Relevant benefit-plan changes have included enhanced coverage beginning in 1986 for treatment of motor vehicle accident injuries if the covered individual was wearing a seat belt at the time of the accident, and merging of sick days with vacation days in 1985 to offer an employee-managed "Personal Days Off" program that provides an incentive to maintain good health.

Other company policies also have had an impact on risks. Cost-containment issues are communicated to all individuals in the organization through a variety of channels and media. This focus on communication helps to ensure widespread understanding of programs and the incentives built into them. The StayWell newsletter, *WellTimes* is also distributed to all employees to keep them abreast of wellness issues. Active recreation clubs, accessible to most employees, have been effective in involving many employees in exercise activities. A corporate smoking policy restricts smoking at company work sites, and cigarette machines were recently removed from all company facilities.

These programs and benefit changes have worked in concert with the Stay-Well program to provide additional incentives for making healthy lifestyle changes. The net effect of these forces can be seen by comparing Employee Health Survey results from 1980 to 1986. Employee Health Survey data are used to assess overall corporate risk and to compare work sites where the StayWell program is fully in place with work sites not yet fully implemented (Anderson & Jose, 1987; Jose & Anderson, 1986; Jose, Anderson, & Haight, 1987).

Risk in two key areas, smoking and weight, has been reduced to a significantly greater degree at sites where the entire StayWell program is available, compared with other Control Data work sites where the StayWell program has not yet been fully implemented (Figs. 5.5 and 5.6). Overall, the percentage of Control Data employees who smoke has decreased from 35% in 1980 to 26% in 1986. However, the 24% smoking rate at StayWell sites is significantly lower than the 30% at other Control Data work sites after adjusting for age, gender, and educational differences ($p < .05$). The percentage of employees who are overweight has not declined significantly company-wide. However, the significant difference between StayWell sites and other Control Data sites after adjusting for age, gender, and educational differences ($p < .01$) indicates that the program has been effective in helping employees to control their weight. On two other risk factors measured by the Employee Health Survey,

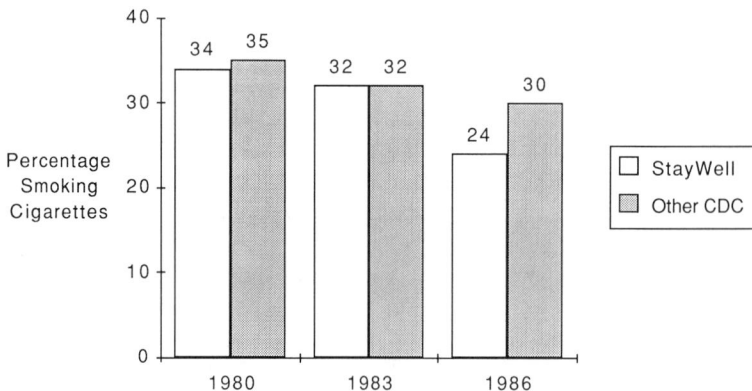

FIG. 5.5. Smoking trends. Source: CDC Employee Health Survey, 1980, 1983, 1986.

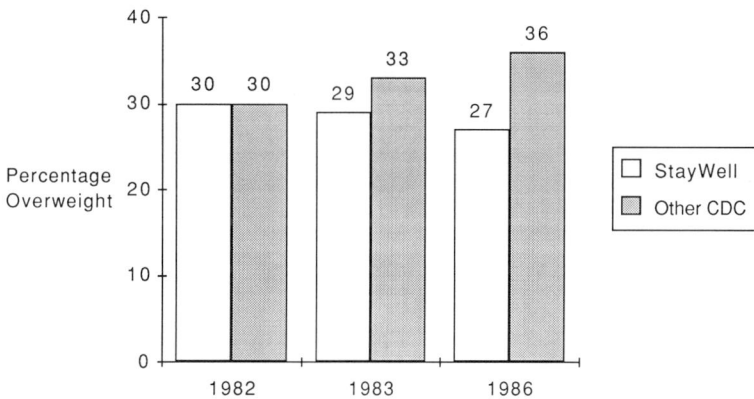

FIG. 5.6. Weight trends. Source: CDC Employee Health Survey, 1982, 1983, 1986.

fitness and seat-belt use, a significant difference in trends was not found between StayWell sites and other Control Data sites when risk levels were adjusted for age, gender, and education. However, there was a 32% company-wide reduction in the proportion of employees who are sedentary, and a 47% drop in individuals not using seat belts.

Fitness and seat-belt use levels among Control Data employees were also compared with levels for external Employee Health Survey customers administered in 1986. These customers included 12 organizations representing 18,498 employees. Results of this comparison indicate that Control Data has significantly higher levels, than this reference group of both fitness ($p < .05$) and seat belt use ($p < .001$) when the effect of age gender, education, and state seat-belt laws are controlled (Figs. 5.7 and 5.8). Controlling for seat-belt laws was necessary since 9 of the 12 customer organizations are in states with seat-belt laws, whereas Control Data employees are concentrated mostly in states where there was no seat-belt law at the time of the survey. These significant differences between Control Data and this external reference group suggest that the corporate-wide elements of the StayWell program (e.g., *Well-Times,* promotions) and other health-promoting company programs (e.g., seat-belt use incentive) are having positive effects.

We are pursuing more detailed analyses in each risk area. Our initial efforts were aimed at the smoking variable since it is known to be associated with high costs. Many more people quit smoking at StayWell sites than actually took the StayWell "How to Quit Smoking" course during the measurement period. Analysis indicated that nearly 1,200 more employees quit smoking at StayWell sites than would have quit if the program had not been at those sites.

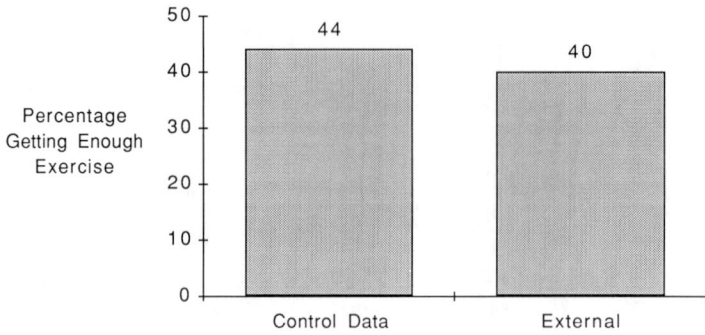

FIG. 5.7. Fitness levels. Control Data vs. External Customers (Adjusted for Demographic Differences). Employee Health Surveys, 1982–1986.

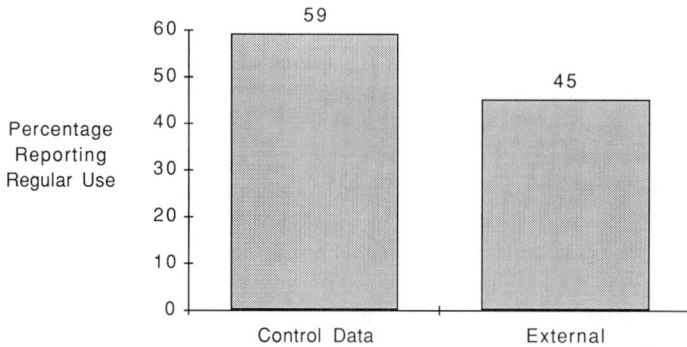

FIG. 5.8. Seatbelt use levels. Control Data vs. External Customers (Adjusted for Demographics/State Laws). Source: Employee Health Survey; 1982–1986.

Participation records indicated that only 433 employees completed the "How to Quit Smoking" course at these sites. Thus, the smoking course itself can account for only a fraction of the employees who quit smoking at StayWell sites. The remainder may be attributable to a "spillover" effect of participation in other StayWell activities and work-site culture changes created by the StayWell program.

Data on all employees who completed two or more health-risk profiles were examined. Those who were smokers at the time of their first health-risk profile were analyzed to determine whether they had quit smoking at the time of their most recent health-risk profile. The mean time between measurements was 28 months. Assuming a quit rate in the general population in the absence of specific interventions of 4% per year (Wilbur, 1986), the number of smokers in our sample at the time of the follow-up health-risk profile should have decreased by 10%. In fact, we observed a 23% overall decrease in smokers. Furthermore, the quit rate was 29% among those who completed StayWell smoking-cessation activities, while the quit rate for smokers not completing such activities was 22% ($p < .07$). Analysis also indicated that activity completers were heavier smokers at the baseline, and that overall reduction in number of cigarettes smoked was 57% greater than the reduction by those who did not complete smoking activities ($p < .05$).

This analysis leads us to conclude that the StayWell smoking-cessation activities are effective. The analysis also supports the notion that the higher-than-expected overall smoking quit rate at StayWell sites reflects the pervasive impact of a comprehensive health-promotion program, beyond that produced by stand-alone courses.

Cost of Risks

One reason risk reduction is of interest to an employer is that it is assumed to lead to reduced health-care claims and increased productivity. Claims analysis by diagnosis groups for health-care claims paid by Control Data in 1984 indicates that 48.5% of the dollars paid were for lifestyle-related categories. Health-promotion activities focus on reducing the cost of this sizable portion of health-care expenses. The StayWell evaluation has shown that high-risk individuals have substantially higher health-care costs and absenteeism due to illness than do low-risk individuals.

A recent study of 15,000 Control Data employees was a significant milestone in quantifying the relationship between health-care costs and employee lifestyle (Anderson & Jose, 1987; *Health Risks and Behavior*, 1987). The study was conducted jointly by Control Data and Milliman & Robertson, Inc., a highly respected actuarial consulting firm specializing in the health-care field. Results provide strong support for one of the major premises of health promotion by demonstrating conclusively that

. . . a significant difference exists in the utilization and cost of medical care by health risk status. Generally, high-risk persons utilize more medical care than other persons and generate higher claim costs. (*Health Risks and Behavior*, 1987, p. 5)

Seven lifestyle risks were examined: exercise, weight, smoking, hypertension, alcohol use, cholesterol, and seat-belt use. Risk levels on these seven factors were linked to three key indicators of health-care costs: non-maternity claims costs, hospital inpatient days per 1,000 employees, and percentage of employees with claims over $5,000. The study details the excess health-care utilization and costs incurred by employees at risk for each of these factors. For example:

- Employees who smoke a pack of cigarettes per day have 18% higher medical claim costs than those who do not smoke.
- Sedentary employees have 30% more hospital days than those who get adequate levels of exercise.
- Seriously overweight employees are 48% more likely to have claims exceeding $5,000 during a one-year period than those at normal weight levels.

Anomalies were found in the relationship between risk and costs for alcohol use and cholesterol. With the exception of these two factors, however, the general finding across risk areas for each of the three indicators of health-care costs was that high-risk employees have higher health-care utilization and costs than low-risk employees.

The widespread response to this research (Controlling High-Risk Behavior, 1987; Negotiating Lifestyle, 1987; Study Lays Groundwork, *Medical Benefits*, 1987; Study Lays Groundwork, *Wall Street Journal*, 1987) probably stems from the scope of the analysis and from the reliability and representativeness of the results. Statistical analysis performed on a number of dimensions by Milliman & Robertson indicates that the data are highly reliable and representative of a nationwide population.

A study by Control Data reached the same general conclusion regarding the relationship between health risks and absenteeism due to illness (Jose & Anderson, 1986). This study looked at random samples of the employees of 31 organizations. A total of 5,787 employees were studied. Employees were classified as high, moderate, or low risk as in the previous study, and six risk factors were measured: smoking, weight, nutrition, exercise, stress, and seat-belt use.

With the exception of nutrition, high-risk individuals were absent due to illness more than low-risk individuals. For example, sedentary individuals were 20% more likely to be absent from work over a week per year due to illness than active individuals. Similarly, smokers were 43% more likely to be absent over a week than were nonsmokers, and those at high risk on the stress factor were 54% more likely to be absent over a week than those at low risk. (Fig. 5.9)

Another compelling finding indicated that the number of factors on which an individual is at risk is also a significant indicator of absenteeism. For those employees not at risk on any of the six factors, only 3% were absent more than a week due to illness. In contrast, 12% of employees at risk on three factors were absent more than a week due to illness, and fully 21% of employees at risk on six or seven factors were absent more than a week due to illness (see Fig. 5.10).

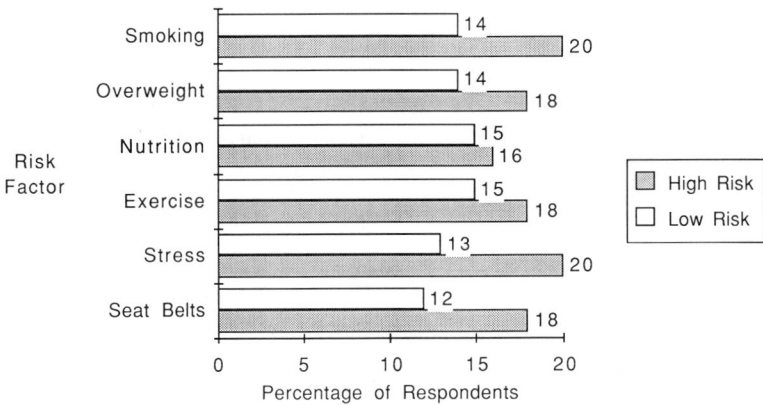

FIG 5.9. Employees absent over five days during the past year due to illness (*N* = 5,787). Source: Employee Health Survey Customer Data, 1982–1986.

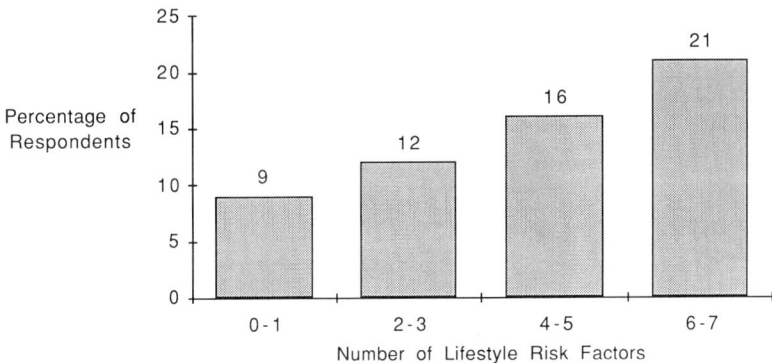

FIG. 5.10. Employees absent over five days during the past year due to illness (*N* = 5,787). Source: Employee Health Survey Customer Data, 1982–1986.

COST SAVINGS ANALYSIS

Modeling the Costs of Risk Factors

There are many employer costs associated with lifestyle risks. The StayWell evaluation has focused on two major types of costs associated with lifestyle risks—health-care claims costs and absence from work due to illness. These costs were associated with six major lifestyle risk factors—smoking, exercise, overweight, blood pressure, cholesterol, and seat-belt use. Multiple linear regression was used to create mathematical models that predict health-care costs and absenteeism based on these risks, controlling for age and job category. Separate models were developed for males and females. Health-care claims models were based on data from 8,286 males and 6,981 females; absenteeism models were based on 9,959 males and 8,040 females. We will provide several examples of the association between lifestyle risks and health-care claims and absenteeism. All of the dollar amounts reported here are 1984 dollars.

Figure 5.11 compares annual projected health-care claims for a high-risk versus a low-risk male at various ages. The high-risk male was defined to be one who smoked two packs of cigarettes per day, was sedentary, was 30% overweight, and did not use seat belts. The low-risk male was a nonsmoker who exercised moderately, was not overweight, and wore seat belts most of the time. It should be noted that although the absolute dollar difference between the high- and low-risk groups increases with age, the percentage difference declines. Figure 5.12 displays the same data for females. Costs for

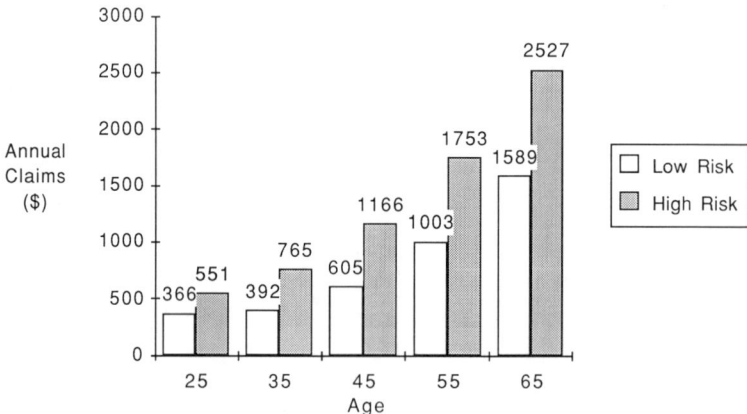

FIG. 5.11. Annual healthcare claims. Low- vs. High-risk* Males (1984 Dollars)

*Smokes 2 packs-day; sedentary; 30% overweight; no seat belts.

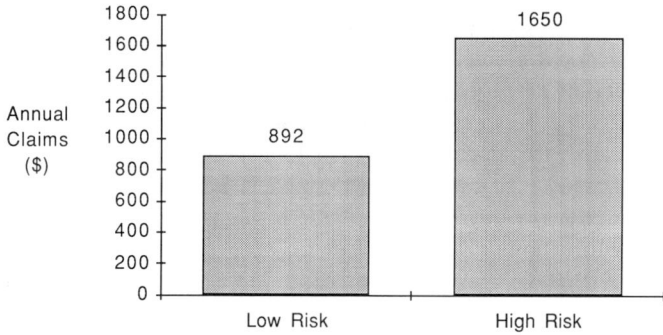

FIG. 5.12. Annual healthcare claims. Low- vs. High-risk* Females (1984 Dollars).

*Smokes 2 packs-day; sedentary; 30% overweight; no seat belts.

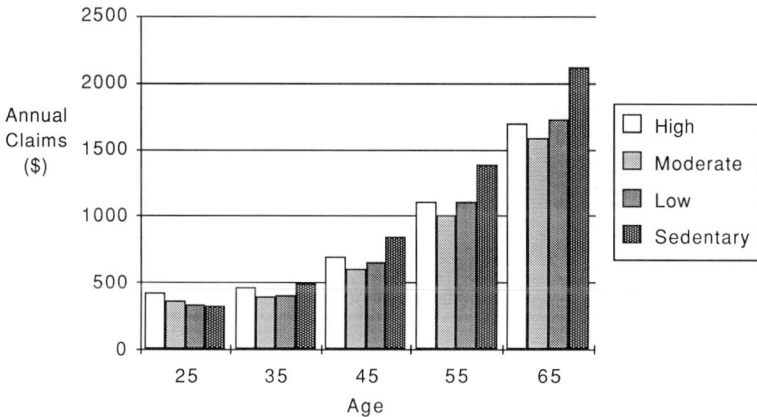

FIG. 5.13. Annual health care claims for males by age and exercise level (other risks at minimum).

females do not vary significantly with age, due to gynecological and obstetric costs in younger women.

The relationship between exercise and male health-care claims costs is examined in Figure 5.13. The dollar amounts given are for males with no other health risks at various levels of exercise. Two aspects of these data are very interesting. First, for the youngest age group, costs increase with exercise, while for the other age groups, costs decrease with increasing exercise. Second, for all age groups above the lowest, the minimum cost is associated with the second highest level of exercise. It is believed that both of these phenomena may be due to the effects of exercise-related injuries. Future research will test this hypothesis.

Figure 5.14 shows data relating exercise levels to claim costs for females. Again, age is not a predictive variable for female health claims. The figure shows that lack of exercise is related to higher claim levels.

The relationship between smoking and health-care claims costs for males and females are examined in Figures 5.15 and 5.16 respectively. All other risks are assumed to be minimal, and smokers were assumed to smoke two packs of cigarettes per day. For males, the percentage increase in claims costs associated with smoking decreases with age, although the actual dollar amounts associated with smoking increase. Since age is not a factor for fe-

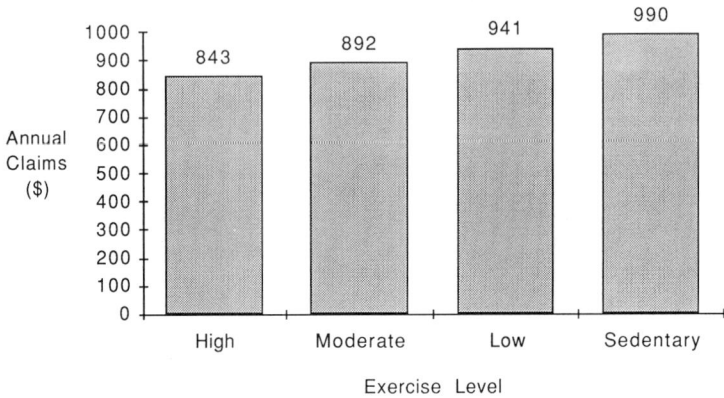

FIG. 5.14.   Annual healthcare claims: females by exercise level.

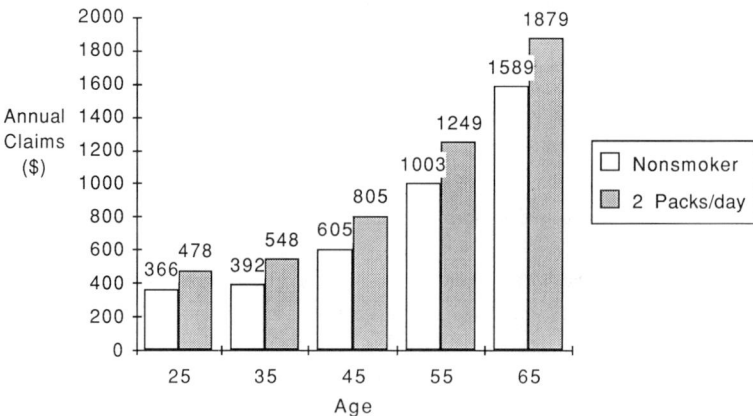

FIG. 5.15.   Annual health care claims for males by age and smoking level (other risks at minimum).

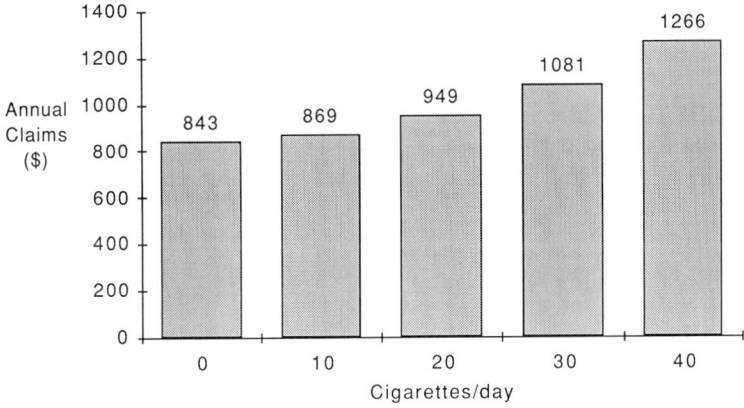

FIG. 5.16. Annual health care claims for females by smoking level (other risks at minimum).

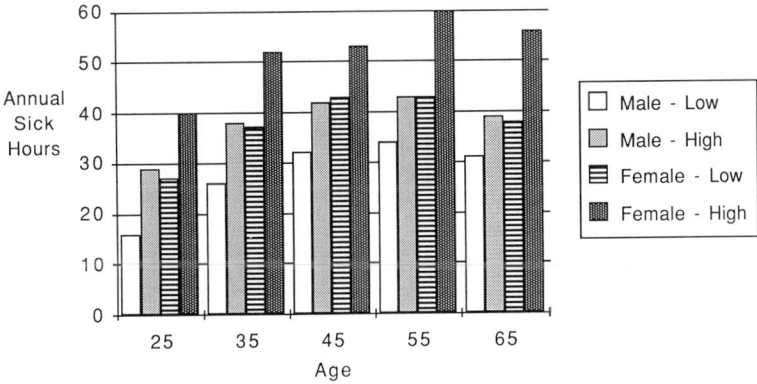

FIG. 5.17. Absenteeism. Low vs. Highrisk*
*Smokes 2 packs-day; sedentary; 30% overweight; no seat belts.

males, expected claims costs at various levels of smoking are shown. For females, claims costs appear to increase exponentially with increasing smoking levels.

The association between several simultaneous risks and absenteeism is illustrated in Figure 5.17. High-risk individuals are predicted to be absent more than low-risk individuals for both genders and at all ages.

Cost Impact of Risk Reduction

Combining excess cost of lifestyle risk from the models with information about change in employee risk over the period from 1980 to 1986, it is possible to estimate Control Data's savings in health-care claims and absenteeism due to risk reduction. This analysis leads to an estimated annual savings in health-care claims alone of approximately $1.8 million.

Although the ability to make such savings estimates is important, it is insufficient for the purpose of determining whether the cost of a wellness program is less than the benefit derived from it. Many other factors need to be considered, such as annual inflation in medical costs, the age at which individuals reduce their risks, and the probability that employees will leave the company in the future (i.e., turnover).

Figure 5.18 summarizes an analysis that accounts for these factors and projects the expected health-care claims both for an individual with minimum risks and one with the same set of high risks used in previous analyses. The expected annual claims are computed from the present until age 65 and then discounted to determine a net present value in 1986 dollars. The cumulative difference between the high- and low-risk individuals is then computed to determine the expected present-value savings due to risk reduction reported in the figure. It was assumed that savings from risk reduction would phase in over four years, with only 25% of the difference between the high- and low-risk individuals being recognized as savings in the first year, increasing to 50% and 75% in the second and third years, respectively. Beginning in the fourth year, the risk-reduction savings were assumed to be 100% of the difference. Also, the turnover data used in this analysis were age dependent based on Control Data's experience. Finally, since the age at which risk reduction oc-

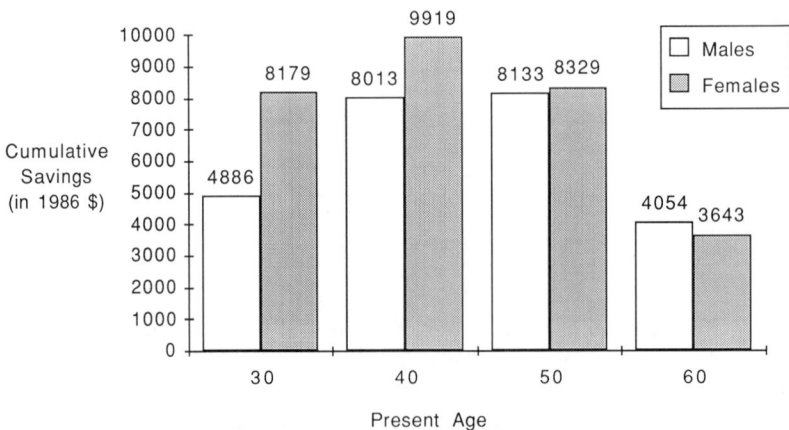

FIG. 5.18. Expected presentvalue savings from reducing several risks (from present age to age 65).

curs has an effect on the total savings to age 65, the expected savings were computed for various starting ages. Combining the results of this type of analysis with data on the effectiveness and cost of the various risk-reduction programs will make it possible to compute the long-term on investment for the StayWell program.

CONCLUSION

Back in the "old days" of the late 1970s and early 1980s, organizations that got into health-promotion programming did so because it seemed like a good idea or because they "believed" in it. In fact, many successful programs were initiated on that basis. Times have changed, however. Purchasers or initiators of health-promotion programs are now almost always forced to confront the economic consequences of such decisions. Health promotion has come to be considered not so much an employee benefit as a crucial cost-management strategy. As such, the economic consequences have become a central concern in the decision-making process.

Evaluation of the StayWell program has demonstrated that individuals who lead healthy lifestyles have lower health-care claims and are absent from work less often than employees with lifestyle risks. Employees of both genders and all ages, representing all job families at Control Data, participate in StayWell activities relevant to their risk factors and make positive lifestyle changes. Together, these results provide evidence that the StayWell program is producing a favorable return on investment for Control Data.

Future evaluation will focus on quantifying this return more precisely by linking risk reduction directly to health-care cost reduction and these bottom-line consequences. The ultimate goal of the evaluation is to produce a model that will enable organizations to plan health-promotion programs that maximize cost effectiveness. The company that ignores the healthy investment in human capital afforded health-promotion programs will operate at an increasing disadvantage in the marketplace as competitors realize their benefits.

REFERENCES

Anderson, D. R., & Jose, W. S. (1987). *Comprehensive evaluation of a worksite health promotion program: The StayWell program at Control Data.* In S. A. Klarreich (Ed.), *Health and fitness in the workplace: Health education in business organizations* (pp. 284–298). New York: Preaeger.

Anderson, D. R., & Jose, W. S. (1987). Employee lifestyle and the bottom line: Results from the StayWell evaluation. *Fitness in Business, 2,* 86–91.

Controlling high-risk behavior can lower health costs: Study. (1987, April 20). *AHA News.*

Cook, T. D., & Campbell, D. T. (Eds.). (1979). *Quasi-experimentation: Design and analysis issues for field settings.* Chicago: Rand McNally.

Fielding, J. E. (1984). Evaluation of worksite health promotion: Some unresolved issues and opportunities. *Corporate Commentary* (Washington Business Group on Health), *1*, 9–15.

Haight, S. A. (1985). StayWell program evaluation: Database description and analysis plans. *Proceedings of the 21st Annual Meeting of the Society of Prospective Medicine*, 79.

Health risks and behavior: The impact on medical costs. (1987). Milwaukee: Milliman & Robertson, Inc.

Jose, W. S., & Anderson, D. R. (1986). Control Data: The StayWell program. *Corporate Commentary* (Washington Business Group on Health), *2*, 1–13.

Jose, W. S., Anderson, D. R., & Haight, S. A. (1987). The StayWell strategy for health care cost containment. In J. Opatz (Ed.), *Health promotion evaluation: Measuring the organizational impact* (pp. 15–34). Stevens Point, WI: National Wellness Institute.

Jose, W. S., Anderson, D. R., & Peterson, K. E. (1985). The Employee Health Survey: A complement to health risk appraisal. *Proceedings of the 21st Annual Meeting of the Society of Prospective Medicine*, 76–78.

Jose, W. S., Williams, W. R., & Keller, P. E. (1986). StayWell program improves employee health and vitality at Control Data Corporation. *Wellness Management* (National Wellness Institute), *2*(2), 1–4.

Koop, C. E. (1983). *The health consequences of smoking: Cardiovascular disease—a report of the Surgeon General.* Rockville, MD: U.S. Department of Health and Human Services.

Negotiating lifestyle: An employer trend. (1987, May 20). *Hospitals*, p. 60.

Study lays groundwork for tying health care costs to workers' behavior. (1987). *Medical Benefits*, *4*, 1–2.

Study lays groundwork for tying health care costs to workers' behavior. (1987, April 14). *The Wall Street Journal*, p. 35.

Wilbur, C. S. (1986, April). Wellness: An introduction. Presentation to Risk Insurance Management Society, Toronto, Ontario, Canada.

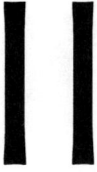

II TARGETS FOR
HEALTH PROMOTION

6 Cigarette Smoking at Work: Data, Issues, and Models

Neil E. Grunberg
Department of Medical Psychology
Uniformed Services University of the Health Sciences

INTRODUCTION

Cigarette smoking at the work site has recently become the focus of public and legislative interest. Private businesses, the Federal government, and local municipalities are rushing to establish no-smoking areas in public places and work places or to pass ordinances restricting smoking in public places and work places. Some work sites are offering smoking-cessation programs free of charge. The intentions motivating this latest anti-smoking movement are certainly good: to encourage smokers to quit and to clean up the environment inside closed spaces. But is this crusade based on a sound data base and real risks? Moreover, can smoking cessation be legislated by fiat? Also, is state-of-the-art knowledge about smoking behavior and the most effective cessation techniques being considered and included in the restriction and control of smoking at the work site?

This chapter will address these questions by reviewing the data bases for smoking cessation at the work site and the types of programs used at the work site. Then, recent developments in smoking research that may be relevant to smoking cessation at the work site will be discussed.

The opinions or assertions contained herein are the private ones of the author and are not to be construed as official or reflecting the views of the Department of Defense or the Uniformed Services University of the Health Sciences.

SMOKING AT THE WORK SITE: HEALTH RISKS

There are several issues that should be considered when discussing smoking at the work site. Most broadly, these issues can be classified into two categories: (1) the health risks of smoking at the work site, and (2) the role of the work site in health promotion. The health risks of smoking at the work site can be divided into risks to the smoker and risks to the nonsmoker. The role of the work site in health promotion in this context involves smoking-cessation programs and policies that restrict or ban smoking.

Health Risks of Smoking to Smokers at Work

The risks of smoking for the smoker are now well-known and extensively documented. The Reports of the Surgeon General on Smoking and Health provide a mountain of convincing data that smoking is the single most preventable cause of death and illness, accounting for a staggering number of cases of cardiovascular disease, cancers, and chronic obstructive lung diseases each year (USDHEW 1979; USDHHS, 1982, 1983, 1989). It is relevant to the present discussion to note how recently the health risks of smoking have become clear. This perspective may help to explain why the medical community and public as a whole have recently turned against smoking with such fervor and have adopted sentiments that are somewhat reminiscent of attitudes that led to alcohol prohibition during the 1930s.

As recently as 1960, cigarette smoking was not considered to be a major health hazard by the public or by the majority of the medical community. In fact, the first Surgeon General's Report on Smoking and Health issued in 1964 (USPHS, 1964) was greeted with widespread skepticism. Basically, the 1964 Report said that smoking might be hazardous to one's health, but it provided relatively little evidence to support even this weak assertion. This report, however, was a watershed document in that it provided the impetus for careful attention to and collection of relevant medical and scientific data. In fact, the 1964 Report was a harbinger of the forthcoming tidal wave of evidence against smoking. But in 1964, the case against smoking was weak, the tobacco industry was hardly tarnished, and tobacco industry stock was still an excellent investment.

The 1967 Surgeon General's Report on Smoking and Health (USPHS, 1968) took a stronger position regarding the health hazards of smoking, and, largely as a result of this report, warning labels were mandated by Congress to appear on cigarette packs. It was not until the landmark 1979 Surgeon General's Report (USDHEW, 1979), however, that a massive and unequivocal body of damning evidence against tobacco use was compiled in one document. In addition, the 1979 Report was the first such report to present thorough reviews of the behavioral and psychosocial variables that are involved in the initiation and maintenance of tobacco use and that must be considered in order to design effective prevention and cessation programs or techniques.

Not until the early 1980s—as the 1979 Report findings became well-known and the 1982, 1983, and 1984 Reports provided more detail about the causative role of smoking in cancer, cardiovascular disease, and chronic obstructive lung diseases—did the risks of smoking become accepted as fact in the eyes of the general public. With this shift in perception also came a shift in public and work site policies. These policies are certainly intended to promote health and to prevent disease. It is unquestionable that smoking poses health risks to the smoker and, therefore, is dangerous for the smoker at the work site. Unfortunately, some of the approaches designed to discourage or prevent smoking reflect a single-minded attack on a health-damaging behavior without fully considering why people use tobacco, and therefore, what may happen if tobacco use is prevented by rules without accompanying cessation programs.

In addition to the risks of smoking per se to smokers, there may be interactions of smoking with other work exposures that potentiate the risks of smoking to the smoker. For example, the presence of flammable substances poses additional risks if a lit match or burning cigarette is nearby. Moreover, some work site materials can cause synergistic health-damaging effects in the smoker. Probably the best example of this type of potentiation is asbestos exposure. Asbestos workers who are nonsmokers have a five times greater risk of developing lung cancer than do nonsmokers who do not work with asbestos. Smokers who are not asbestos workers have a tenfold greater risk of developing lung cancer compared to nonsmokers who are not asbestos workers. Smokers who also are asbestos workers have a 50-fold greater risk of lung cancer compared to nonsmokers who do not work with asbestos. That is, there is a multiplicative relationship between smoking and asbestos exposure. There are definite health risks of smoking for the smoker at work from smoking per se and from interactions of smoking with exposure to other materials.

Health Risks of Smoking to Nonsmokers at Work

What about the risks of smoking to nonsmokers at work? There certainly are risks to everyone in the vicinity of a lit match or burning cigarette when flammable substances are nearby. But this type of risk is rather rare because everyone is highly motivated to keep flames away from flammable materials, and smokers in settings with flammable materials either smoke elsewhere at the work site or use smokeless tobacco products, such as chewing tobacco.

The major question regarding the health risks of smoking to the nonsmoker at the work site refers to potential dangers of "passive smoking," also called "involuntary smoking" or "environmental tobacco smoke" (ETS). This topic has received substantial media attention as a result of the 1986 Surgeon General's Report (USDHHS, 1986). The major conclusions of that report were:

1. Involuntary smoking is a cause of disease, including lung cancer, in healthy nonsmokers.

2. The children of parents who smoke compared with the children of nonsmoking parents have an increased frequency of respiratory infections, increased respiratory symptoms, and slightly smaller rates of increase in lung function as the lung matures.

3. The simple separation of smokers and nonsmokers within the same air space may reduce, but does not eliminate, the exposure of nonsmokers to environmental tobacco smoke. (p. 7)

These conclusions have been combined and paraphrased by the media and by public-policy makers into the simple generalization that involuntary smoking poses a serious health risk to nonsmokers and that the only way to prevent this risk is to ban smoking in public places (including work sites). This sweeping generalization oversteps the data presented in the 1986 Report and is an inaccurate oversimplification of the findings of that report. A careful reading of the 1986 Report reveals that involuntary smoking certainly raises the relative risk of some pulmonary diseases in some studies of long-term, chronic exposure of nonsmokers to environmental tobacco smoke. Even in studies where an effect was detected, however, the effect size was rather small and the small effects were detected with epidemiological relative risk calculations. Also, the nonsmokers who clearly showed increased pulmonary diseases and dysfunction were spouses or children of smokers exposed constantly or for many years to tobacco smoke.

With regard to the health risks of work-site smoking exposure to nonsmokers, the 1986 Report wrote:

> The workplace, an important source of tobacco smoke exposure, was not considered in the early studies of involuntary smoking. Later case-control studies provided some information on tobacco exposure at work, but the data were limited and inconclusive. (p.91)

The report goes on to cite five studies related to health risks of environmental tobacco smoke to nonsmokers at work. Two of these five studies reported relationships between work-site tobacco smoke exposure and lung cancer (Kabat & Wynder, 1984; Koo, Ho, & Saw, 1984), one study reported a nonsignificant, relative risk trend (Wu, Henderson, Pike, & Yu, 1985), and two studies reported no relationship (Garfinkel, Auerbach, & Joubert, 1985; Lee, Chamberlain, & Alderson, 1986). Even the two studies that reported relationships require caveats. The Kabat and Wynder (1984) study found a relationship only for men at work and not for women, and this same study found no health risks among nonsmoking men or women with a spouse who smoked. In addition, type of work was not considered in the data analyses as a potentiating or confounding variable in this study. Koo et al. (1984) was based on only two cases and four controls.

With regard to involuntary smoking at the work place, so far there is relatively little direct evidence that a risk is posed, in general. It remains possible that this lack of evidence may reflect the lack of studies on this topic and that current attention to ETS soon will provide new data. It also may be that environmental tobacco smoke poses risks at some work sites and not at others. For example, if the work site is constantly filled with a great deal of smoke, such as in some restaurants, the nonsmoking worker may be at elevated risk for some diseases. However, this type of connection is speculative at this time and does not currently have a supporting data base. Also, high ceilings, exhaust fans, and good ventilation probably can offset the risks posed by environmental tobacco smoke at most work sites. At present, based on the available data, health risk to nonsmokers is not a convincing reason for smoking cessation at the work site. However, tobacco smoke does introduce toxins into the environment and as such may require regulation.

The Role of the Work Site in Health Promotion

What is the work site's role in health promotion? More specifically, what is the work site's role in smoking cessation? What types of smoking cessation programs are provided? Do they do any good? Should policies be established to restrict or ban smoking? Do such policies help the smoker to give up tobacco? Is anyone helped by these programs and policies? Is anyone hurt by them? Are there better ways to encourage smoking cessation at work? The rest of this chapter reviews and discusses data bases and current research developments relevant to these questions.

STATE OF KNOWLEDGE AND DATA BASE ON SMOKING CESSATION AT WORK

Fielding (1982) wrote that "controlled experimental data on the efficacy of such work site (smoking reduction) programs is virtually nonexistent" (p. 909). This sorry state of affairs has improved some since 1982, but even today there are few experimental evaluations of work-site smoking-cessation programs. This is not surprising because it is difficult to do health-promotion interventions at work sites that include random subject assignment to experimental and control groups. Fortunately, there is now a fair number of relevant studies in the literature, some of which are true experiments. These studies have been critically evaluated in the 1985 Surgeon General's Report and in more recent review papers.

1985 Surgeon General's Report

The 1985 Surgeon General's Report (USDHHS, 1985) had two major conclusions:

1. For the majority of American workers who smoke, cigarette smoking represents a greater cause of death and disability than their workplace environment.

2. In those worksites where well-established disease outcomes occur, smoking control and reduction in exposure to hazardous agents are effective, compatible, and occasionally synergistic approaches to the reduction of disease risk for the individual worker. (p. 11).

Within this report, Chapter 12, entitled "Smoking Intervention Programs in the Workplace," is particularly relevant to the present discussion. Therefore, the findings of that chapter are reviewed in some detail in this section.

The Cost of Employees Who Smoke

Overall, smokers are estimated to use the health-care system as much as 50% more than do nonsmokers (Fielding, 1984). In addition, smokers have higher rates of work-related accidents, disability, reimbursement payment, and absenteeism than do nonsmokers (Terry, 1971). The cost of the smoking employee to the employer can be great. (See chapters by Fielding and by Johnson in this volume.) A smoking-cessation program, therefore, could be cost-effective as well as health-promoting.

Major Types of Smoking-Cessation Programs at Work

According to the 1985 Report, the smoking modification approaches being offered at work sites can be broken down into six categories: (1) monetary incentives, (2) contests for not smoking, (3) distribution of self-help materials, (4) physician messages, (5) health-education lectures, and (6) stop-smoking clinics. There are advantages and disadvantages to these work site smoking-cessation programs. In general, advantages include: convenient access; reduced cost (assuming the company pays); friends and co-workers are the co-participants (rather than strangers); and participants learn not to smoke in the context and environment where they spend much of their time, in contrast to trying to carry a program from an outside clinic into work where environmental cues and situations are different.

But there also are disadvantages to work site smoking-cessation programs. The times of meetings may be inconvenient and could interfere with or interrupt work schedules. Then, the worker must choose whether to deleteriously affect ongoing work or to skip the meeting. Another problem is that there may be either coercion or a perception of coercion to participate in a work site program. This situation could create reactance, alter perceived control and choice, or in other ways undermine the influence of an otherwise effective program. A third problem is that nonsmokers may come to resent the "time off" given to the smokers. Also, if there are other health risks at the job, such as asbestos exposure, unions and individual employees may interpret the program as a way to decrease the employer's liabilities and responsibilities to deal with those health risks. Providing a smoking-cessation program at work can raise difficult legal as well as perceptual issues.

Criteria for Evaluating These Programs

The criteria for evaluating the programs' effects include: (1) changes in participants' smoking behavior, (2) effects on smoking and health-related variables for all employees, and (3) secondary effects of the program on nonhealth variables of concern to the employers. Changes in participants' smoking behavior poses the standard problem of smoking or abstinence verification. Self-reports are the least expensive approach to gather this information, but the validity of self-reports of smoking behavior is questionable. The best smoking-cessation intervention trials today use some sort of biochemical analysis (such as salivary continine) to determine whether subjects are still smoking and to gather some indirect quantitative measure. This type of sophisticated measure has not been included in most work-site smoking-cessation program evaluations. The second criterion basically involves a control group. To know whether a given program is successful, it is important to compare participants' smoking behavior to a matched group of nonparticipants. Third, it may be useful and meaningful to determine whether a program affects morale or turnover ("secondary effects") of the participants or other employees. These three evaluation criteria should be considered in reviewing any work-site smoking-cessation program.

Work-Site Smoking-Cessation Programs

Many of the studies of work-site smoking-cessation programs reviewed in the 1985 Report were uncontrolled. Table 6.1 summarizes 13 such studies considered in the 1985 Surgeon General's Report. The participants in these studies were self-selected; abstinence was determined by self-report when it was reported; and no control groups were included in these studies. The self-reports of abstinence are particularly questionable because some studies excluded subjects who dropped out of the program and because several studies provided financial rewards for self-reports of smoking abstinence. Overall, these studies provide little, if any, meaningful information.

There have also been controlled studies of work-site smoking-cessation programs (see Tables 6.2 and 6.3). Overall, the 6–24 month follow-up abstinence

TABLE 6.1
Uncontrolled Studies Without Objective Measures of Smoking Status

Study	Number of subjects, type of worksite	Cessation rate (percent)	
		Posttreatment	Followup (No. months)
Andrews (1983)	965 hospital employees	Not reported	26 (20)
Bauer (1978)	81 Bell Laboratories employees	90	30 (6)
Bishop and Fisher (1984)	10–46 employees in each of six companies	25–60	6.5–33 (12)
Dawley et al. (1984)	15 VA hospital employees and 2 patients	88	50 (6)
Ellis (1980)	Asbestos company employees	Not reported	30 (48)
Grove et al. (1979)	33 Blue Cross employees	33	27 (6)
Heckler (1980)	16 Thomas Lipton, Inc. employees	Not reported	50 (1)
Kanzler et al. (1976)	9 psychiatric institute employees and 21 community members	67	40 (12)
Miller (1981)	33 engine manufacturing company employees	Not reported	55 (12)
Rosen and Lichtenstein (1977)	12 ambulance company employees	58	33 (12) (at work)
Shepard (1980)	26 electronics mfg. company employees	Not reported	35 (48) (at work)
Sorman (1979)	55 Riviera Motors employees	Not reported	31 (12)
Stachnik and Stoffelmayr (1983)	Employees in three companies: bank, manufacturer, and health services	Not reported	80–91 (6)

Source: U. S. Department of Health and Human Services (1985), p. 482.

TABLE 6.2
Work-Site, Subject, and Procedural Characteristics of Controlled
Outcome Studies

Study	Size and type of worksite	Participation rate (percent)	Characteristics of participants	Attrition rate (percent)	Recruitment strategies
Abrams et al. (1985)	800-employee medical manufacturing company and 1,600-employee insurance carrier	Not reported (estimated 6)	54 clerical and blue-collar employees	42	Paycheck stuffers, posters, newsletter articles
Glasgow et al. (1984)	600-employee telephone company	Not reported (estimated 18)	25 female, 11 male employees	Not reported	Employee organization sponsorship, newsletter notices, posters
Glasgow et al. (in press)	VA hospital, health care services company, and savings and loan	Not reported	20 female, 9 male employees	7	Brochures, posters, newsletter notices, memos
Klesges et al. (1985)	Four banks and one savings and loan, 115–180 workers each	88 with competition; 53 without (p <0.05)	82 female, 25 male employees	9	Brochures, announcements by bank presidents, time off work for participation; prize to bank with highest participation
Kornitzer, Dramaix et al. (1980)	30 Belgian factories	84 agreed to screening	19,390 male employees, aged 40–59 years; high risk: upper 20 percent of risk distribution	Not reported	Not reported
Li et al. (1984)	Naval shipyard	87	871 male shipyard workers	17	Participation asked at required screening

(Continued)

TABLE 6.2
(Continued)

Study	Size and type of worksite	Participation rate (percent)	Characteristics of participants	Attrition rate (percent)	Recruitment strategies
Malott et al. (1984)	Medical clinic and telephone company	Not reported (estimated 7)	20 female, 4 male employees, primarily clerical and nurses	0	Newsletter notices, brochures distributed by supervisors, recruitment in lunchrooms
Meyer and Henderson (1974)	Varian Corporation; 240 employees, volunteers for risk factor screening (13 percent of workforce)	Not reported	36 employees identified at screening as high risk for cardiovascular disease	0	Invitation to health screening
Nepps (1984)	Johnson & Johnson Corporation	Not reported	36 white-collar employees: 20 women, 16 men	67	Posters, desk drops, company newsletter
Rand et al. (1984)	Large city hospital	Not applicable	18 female employees	Not reported	Advertisements, word of mouth
Rose et al. (1980)	24 large British industrial groups	86 agreed to screening	18,210 male employees, 40–59 years old; high risk: upper 12–15 percent of distribution	6–12	Invitation to health screening exam
Schlegel et al. (1983)	28 Canadian military bases	Not reported	243 armed forces personnel (65 percent male)	Not reported	Posters, news releases
Scott et al. (1983)	Large VA hospital	100	26 nurses (22 women, 4 men)	0 of those continued at VA	Individually approached

Source: U. S. Department of Health and Human Services (1985), pp.484–485.

TABLE 6.3
Design and Outcome of Controlled Work-Site
Smoking-Modification Studies

Study	Program intensity and components	Experimental design	Cessation rate (percent)		Biochemical verification
			Post-treatment	Followup (No. months)	
Abrams et al. (1985)	Basic four-session nicotine-fading cessation program; four-session maintenance treatment	Basic program plus health education (n=18); stress management (n=18); or social support (n=18)	38^1 33^1 6^2	33^1 (3) 27^1 6^2	CO
Glasgow et al. (1984)	Seven weekly small group meetings on brand changing and number reduction; goal choices, abstinence or controlled smoking	Gradual reduction (n=12); abrupt reduction (n=13); gradual plus feedback (n=11)	Not reported	33^1 (6) 0^2 0^2	CO
Glasgow et al. (in press)	(See Glasgow et al. 1984) Social support with two meetings, installments of manual, and phone calls	Basic treatment program (n=13) vs. basic treatment plus significant other social support (n=16)	54 36	25 (6) 23	CO SCN
Klesges et al. (1985)	(See Glasgow et al. 1984) Competition, with monetary prizes, weekly feedback charts	Quasi-experimental; basic treatment (n=16) vs. basic treatment plus competition (n=91)	31 22	14 (6) 18	CO SCN

(Continued)

TABLE 6.3
(Continued)

Study	Program intensity and components	Experimental design	Cessation rate (percent)		Biochemical verification
			Post-treatment	Followup (No. months)	
Kornitzer, Dramaix et al. (1980)	Multiple risk factor program, written advice and antismoking posters; high risk subjects, semiannual physician counseling and stop-smoking booklet	Treatment (n=7,398) vs. screening only (n=8,824)	Not reported	High risk 19[1] (24) 12[2] Random sample 12.5 12.6	No
Li et al. (1984)	One-session physician advice and stop-smoking pamphlet	3- to 5-minute behavioral counseling (n=215) vs. warning to quit (n=361)	Not reported	8.4[1] (3, 11) 3.6[2]	CO
Malott et al. (1984)	(See Glasgow et al. 1984) Coworker support: partner support manual, buddy system, individualized support behaviors	Basic treatment program (n=12) vs. basic treatment plus coworker social support (n=12)	17 17	27 (6) 17	CO
Meyer and Henderson (1974)	Multiple risk factor program, 9 to 12 meetings; 2- to 3.5-hour behavior modification group meetings with spouses	Behavior modification (n=12) vs. individual counseling (n=10) vs. physician advice alone (n=14)	40 25 0	20 (3) 25 33	No
Nepps (1984)	Nine written self-help modules; minimal therapist contact	Quasi-experimental: minimal contact (n=36) compared with earlier group cessation program	22	14 (6)	CO

TABLE 6.3
(Continued)

Study	Program intensity and components	Experimental design	Cessation rate (percent)		Biochemical verification
			Post-treatment	Followup (No. months)	
Rand et al. (1984)	Monetary incentives for low daily CO levels; 1 week of reducing CO levels and 2 weeks of abstinence	Within-subjects design (n=18): baseline—cutdown—abstinence goals	61	28 (3 wks)	CO
Rose et al. (1980)	Multiple risk factor program: posters and stop-smoking booklets; high risk subjects, four company physician consultations	Treatment (n=9,734) vs. screening only (n=8,476)		High risk 12[1] (5 yrs) 0[2] Others 7[1] 0[2]	No
Schlegel et al. (1983)	6-month program; 160-page workbook; abstinence or reduced smoking goal choice; base personnel were therapists	Full treatment (17 sessions) vs. minimal contact (4 sessions) vs. self-help; crossed with nicotine gum/no nicotine gum	45–68[3] (12) 28–31[3] 6–14[3]	25–38[3] (12) 17–29[3] 7–10[3]	No
Scott et al. (1983)	Brief daily sessions with brand fading, treatment manual, and CO feedback, 3 months; abstinence or reduced smoking goal choice	Treatment (n=16) vs. no treatment (n=10)	56[1] 0[2]	25 (9) 0	CO

NOTE: The sample sizes reflect the number of subjects receiving each treatment condition, which in some instances differs from the total number of subjects initiated into the study (see Table 2). Except for the Kleges and colleagues (1985) and the Li and colleagues (1984) studies, in experiments using between-subjects designs there was random assignment to treatment conditions.

NOTE: CO=carbon monoxide; SCN=saliva thiocyanate.

[1,2] At each assessment point, conditions that were significantly different (p <0.05) are identified by different superscripts.

[3] Results of this factorial study are complex and difficult to summarize with a notation system; see text for clarification (numbers for treatment conditions not reported).

Source: U. S. Department of Health and Human Services (1985), pp. 486–488.

rates in these studies look pretty good. However, there are issues to be considered even in these controlled studies: (1) the participants were self-selected, which makes it difficult to evaluate the effectiveness of the programs; (2) greater numbers of sessions led to higher cessation rates; (3) if the work site had more employees, the cessation program was not as effective; (4) the programs specifically focused on smoking cessation were generally more effective than were the multiple risks programs; (5) social support was not particularly effective in these studies (a puzzling finding that contradicts much of the cessation literature); (6) physician advice to quit seemed to increase cessation rates; (7) incentives appeared to be useful; and (8) competitions (between programs, groups, or work sites) increased participation.

Considering all of the available studies up to 1985 of work-site smoking cessation, the Surgeon General's 1985 Report concluded:

1. Smoking modification and maintenance of nonsmoking status among initial quitters has the promise of being more successful in worksite programs than in clinic-based programs. Higher cessation rates in worksite programs are achieved with more intensive programs.

2. Incentives for nonsmoking appear to be associated with higher participation and better success rates. Further research is needed to specify the optimal types of incentive procedures.

3. Success of a worksite smoking program depends upon three primary factors: the characteristics of the intervention program, the characteristics of the organization in which the program is offered, and the interaction between these factors.

4. Research is needed on recruitment strategies and participation rates in worksite smoking programs and on the impact of interventions on the entire workforce of a company.

5. More investigations are needed on worksite characteristics associated with the success of occupational programs and on comprehensive programs including components such as quit-smoking contests, no-smoking policies, physician messages, and self-help materials in addition to smoking cessation clinics.

6. The implementation of broadly based health promotion efforts in the workplace should be encouraged, with smoking interventions representing a major component of the larger effort to improve health through a worksite focus. (pp. 509–510)

Despite the care taken in the 1985 Report to consider all of the relevant data and to reach firm conclusions, it is apparent just how tentative one has to be, based on the dearth of available, convincing data. Conclusions 1 and 2 above are tentative findings: #1 wrote "has the promise of being" and #2 wrote "appear to be associated with." Conclusion #3 is straightforward but rather banal. Conclusions #4 and 5 say that more research is needed to evaluate these programs. And conclusion #6 is a policy statement and not a research finding. Basically, as recently as the 1985 Surgeon General's Report on work-site smoking, there was little convincing research on which to base firm conclusions.

More Recent Reviews of Smoking Cessation at Work

Since the 1985 Report, a few updated reviews have addressed smoking cessation at the work site. Klesges and Glasgow (1986) critically examined uncontrolled and controlled studies of work site smoking cessation and came to similar conclusions as those reached in the 1985 Surgeon General's Report. Based on a review of controlled studies, including their own (Glasgow, Klesges, Godding, Vasey, & O'Neill, 1984; Klesges, Vasey, & Glasgow, 1985; Malott, Glasgow, O'Neill, & Klesges, 1984; Nepps, 1984; Schlegel, Manske, & Shannon, 1983; Scott, Denier, & Prue, 1983), these investigators concluded that: (1) clinic-based smoking-reduction programs can be used in the work site; (2) time off work and involvement of top management in promoting the programs can increase participation rates; (3) absence of negative interactions is associated with treatment success; (4) incentives (e.g., competition as motivator) are important to increase participation; (5) objective verification of smoking is necessary to evaluate program effectiveness; (6) physician stop-smoking messages are useful to include in work-site programs; and (7) controlled studies are necessary for program evaluation.

An even more detailed evaluation of these programs which built upon the 1985 Surgeon General's Report was published by Klesges, Cigrang, and Glasgow (1987). According to this review, the major advantages of work-site smoking-cessation programs for employees include convenience, cost, and friends and co-workers as co-participants. The major disadvantages for the employees include interference with work activities and feelings of coercion to participate. The primary advantages for the employers include increased work productivity, increased employee morale, improved employee and public relations, increased monetary savings from decreased absenteeism, and decreased medical costs. The primary disadvantages for the employers include cost of the program, time off work for participants, resentment by nonsmokers or ex-smoker employees if smoking cessation is the only health-promotion program provided, and perception by the employees that the program is a ploy to decrease health-risk responsibilities and liabilities or to decrease salaries in lieu of these health-promoting fringe benefits.

Klesges, Cigrang, and Glasgow (1987) categorized studies of smoking cessation at the work site into four groups: (1) bibliotherapy or packaged self-help programs; (2) physician advice to stop smoking; (3) multicomponent behaviorally based programs; and (4) programs involving competition or incentives. Of the studies reviewed, few were true experiments, some were quasi-experimental in that work sites were nonrandomly assigned to conditions, and several were uncontrolled with no comparison groups. Seven bibliotherapy and self-help studies were reviewed (Bishop & Fisher, 1984; Dawley, Fleischer, & Dawley, 1984; Ellis, 1980; Heckler, 1980; Jason et al., 1987; Nepps, 1984; Schlegel et al., 1983). These studies were not controlled, they had low cessation rates, they had high attrition rates, but they were cost-effective. The four physician-advice studies (Kornitzer, Dramiaix, Kittel, & DeBacker, 1980; Li et al., 1984; Meyer & Henderson, 1974: Rose, Heller, Pedoe, & Christie, 1980) were well done methodologically, had good participation rates, were low cost, but had modest cessation rates. The behaviorally oriented multicomponent group cessation programs (Abrams et al., 1985; Glasgow et al., 1984; Glasgow, Klesges, & O'Neill, in press; Grove, Reed, & Miller, 1979; Kanzler, Zeidenberg, & Jaffe, 1976; Klesges, Glasgow, Klesges, Morray, & Quale, 1987; Klesges, Vasey, & Glasgow, 1986; Long & Simone, 1985; Malott et al., 1984; Meyer & Henderson, 1974; Norton-Ford & Schmitz, 1982; Schlegel et al., 1983; Scott et al., 1983) were good methodologically (including nine true experiments), had good cessation rates, but had low participation rates. The incentive and competition programs (Klesges, Glasgow, Klesges, Morray, & Quale, 1987; Klesges, Vasey, & Glasgow, 1986; Rand, Stitzer, Bigelow, & Mead, 1984; Rosen & Lichtenstein, 1977; Shepard, 1980; Sorman, 1979; Stachnik & Stoffelmayr, 1983) had high participation rates, good cessation rates, but these tended to be demonstrations rather than experiments.

Based on the available literature, it appears that competition and incentive-based programs attract the most participants and produce good cessation rates. Compared to the other major approaches, competition programs seem to work best at the work site. However, few of these programs have incorporated the most up-to-date knowledge of smoking. Perhaps programs could be improved based on new findings. To consider this possibility, some recent developments in smoking research are briefly reviewed in the next section and are considered with regard to smoking cessation at the work site.

RECENT RESEARCH DEVELOPMENTS ABOUT SMOKING

Despite the health importance of smoking cessation and the potential value to employers as well as employees to encourage smoking cessation, the work-site programs have generally not drawn from current research findings on smoking to design the most effective programs. This section briefly lists some of the active research attention and findings on smoking that might relate to improving work-site smoking-cessation programs.

Tobacco Use as an Addiction

The 1988 Surgeon General's Report (USDHHS, 1988) concluded that:

1. Cigarettes and other forms of tobacco are addicting.
2. Nicotine is the drug in tobacco that causes addiction.
3. The pharmacologic and behavioral processes that determine tobacco addiction are similar to those that determine addiction to drugs such as heroin and cocaine (p. 9).

These conclusions were based on a comprehensive review of relevant data drawn from pharmacology, psychology, behavioral pharmacology, and neurosciences. Based on this extensive review, it is clear that humans and animals display highly controlled or compulsive use of nicotine and tobacco products and that nicotine has psychoactive effects. Cigarette smokers, for example, display drug-reinforced behavior and compensate their smoking behavior to adjust nicotine self-administration. Also, tolerance and dependence develop to tobacco use. In addition, recent studies using nicotine polacrilex gum indicate that this sort of nicotine substitution is a useful adjunct to behavioral programs for smoking cessation (USDHHS, 1988).

In the present context, two points gleaned from the 1988 Report are relevant: (1) smoking cessation is not simply a matter of will power; and (2) nicotine replacement therapy (such as nicotine gum) is a useful pharmacologic adjunct to behavioral smoking-cessation therapy. Few work-site programs have considered these findings and have incorporated these important points. For example, participants in smoking-cessation programs should be told about the addictive properties of nicotine and prepared for the withdrawal phenomena commonly experienced with abstinence. (See Grunberg & Bowen, 1985, for a more detailed discussion of coping with the sequelae of smoking cessation.)

It should be emphasized that many people successfully give up smoking, but that it will be difficult. Co-workers should offer encouragement to the smoker trying to quit and should understand how difficult it may be for the smoker. It is important to point out to smokers who have failed to quit before that it usually takes several attempts to abstain forever and that such failures do not constitute personal weakness. Basically, a realization of the addictive nature of tobacco products can direct attention toward helping the smoker to quit rather than blaming the smoker for smoking.

Effects of Nicotine that Promote Tobacco Use

A related body of data considers effects of nicotine that promote tobacco use but are not traditional dependence effects. These actions include effects of nicotine to enhance human performance, control stress and mood responses, and control body weight. Considering these effects, the 1988 Report concluded:

1. After smoking cigarettes or receiving nicotine, smokers perform better on some cognitive tasks (including sustained attention and selective attention) than they do when deprived of cigarettes or nicotine. However, smoking and nicotine do not improve general learning.

2. Stress increases cigarette consumption among smokers. Further, stress has been identified as a risk factor for initiation of smoking in adolescence.

3. In general, cigarette smokers weigh less (approximately 7 lb less on average) than nonsmokers. Many smokers who quit smoking gain weight.

4. Food intake and probably metabolic factors are involved in the inverse relationship between smoking and body weight. There is evidence that nicotine plays an important role in the relationship between smoking and body weight. (USDHHS, 1988, p. 439)

Few, if any, work-site smoking-cessation programs are seriously considering or addressing these issues. As stated above, it is important to prepare smokers who are about to quit for the sequelae of smoking cessation, including attentional disturbances, sleep disturbances, irritability, anxiety, and weight gain (Grunberg & Bowen, 1985). Such preparation is important to allow people to cope with these unpleasant effects and to avoid relapse (Shumaker & Grunberg, 1986). For example, smokers who quit may find that they cannot think straight or that their attention abilities have been diminished for a little while. It may, therefore, be useful to coordinate quit dates with job demands and with vacation days. Stress-management and weight-control techniques also should be available to smokers preparing to quit smoking. Cessation of smoking leads to real and unwanted effects. Many smoking-cessation programs now include weight-management and stress-management components. These issues also must be addressed in work-site smoking-cessation programs if they are to be effective at helping people to abstain permanently from tobacco use.

Pharmacokinetics and Pharmacodynamics of Nicotine

Pharmacokinetics basically refers to what the body does to a drug (including absorption, distribution, and elimination). Pharmacodynamics refers to what the drug does to the body. Increased sophistication in pharmacokinetic and pharmacodynamic techniques and study designs over the past few years has provided increased information about nicotine presented to the body in various forms (USDHHS, 1988). Most relevant to the present discussion are two points: (1) developments in testing to determine smoking abstinence, and (2) effects of alternative forms of tobacco or nicotine-replacement strategies on nicotine levels in the body.

With regard to testing, measurement of cotinine (the primary metabolite of nicotine) in the saliva or urine provides a good index of tobacco abstinence because it detects smoking for several days after cessation (Benowitz, 1983). Therefore, work-site smoking-cessation intervention trials should use this type of biochemical verification of abstinence to evaluate program success. With regard to different forms of nicotine delivery, programs that recommend or provide nicotine gum should be aware that use of the 2 mg gum currently available in the United States achieves a blood level of nicotine that is roughly 20% of the level of nicotine maintained in many smokers while smoking. That means that chewing the 2 mg gum is not a complete replacement of nicotine but, instead, is a dramatic reduction for most smokers. Too many people do not realize this fact, which partially explains why nicotine gum alone does not help people to quit. However, nicotine gum plus behavioral therapy seems to be helpful for many people (USDHHS), 1988).

Gender Differences in Nicotine's Effects

Another issue that is currently receiving research attention is whether there are gender differences in nicotine's effects and how such differences may affect smoking cessation. Some animal studies indicate greater effects of nicotine on the body weight and eating behavior of females as compared to males (Grunberg, Bowen, & Winders, 1986; Grunberg, Winders, & Popp, 1987). These findings may be related to the fact that women generally smoke fewer cigarettes per day and take fewer puffs per cigarette than do men. Perhaps, the effects of smoking abstinence are quantitatively or qualitatively different in men and women. Currently, the jury is still out on this question. However, men and women commonly report different reasons for smoking and different reasons for refusing to quit (Sorenson & Pechacek, 1987). Therefore, cessation programs probably should consider the gender of the participants and should try to tailor some of the program to the particular concerns of the individuals.

Mechanisms of Nicotine's Actions

It has long been known that nicotine has cholinergic actions in the body, mimicking many effects of acetylcholine on the autonomic nervous system to affect the cardiovascular, respiratory, musculoskeletal, and gastrointestinal systems. Recently, it has become clear that nicotine binds to receptors in the brain. In addition, nicotine affects catecholamines, serotonin, endogenous opioid peptides, steroids, and insulin in the body (USDHHS, 1988).

In light of these widespread effects, it is unlikely that a pharmacologic antagonist to nicotine will soon be developed that blocks effects of nicotine without disturbing the body. However, these new findings about mechanisms of action of nicotine may suggest new approaches to smoking cessation. For example, based on a series of human and animal studies, Grunberg (1986, 1990) postulated that the unpleasantness of nicotine withdrawal might be reduced by consumption of high-carbohydrate foods. Based on this idea and the work of Fernstrom, Wurtman, and Wurtman (Fernstrom & Wurtman, 1971, 1972; Wurtman, 1987; Wurtman & Wurtman, 1979), Bowen, Spring, and Fox (1989) performed a clinical trial in which one group of smokers trying to quit consumed a high-carbohydrate diet supplemented with tryptophan, and another group of smokers trying to quit consumed a high-protein diet. A two-week follow-up indicated that the group on the high-carbohydrate diet reported fewer withdrawal symptoms and were smoking less than the comparison group. It is too early to conclude whether a stop-smoking diet really exists, but this type of extrapolation based on mechanism studies may prove to be clinically significant.

Relapse Prevention

An important aspect of successful, long-term smoking cessation is relapse prevention. As Mark Twain wrote: "To cease smoking is the easiest thing I ever did; I ought to know because I've done it a thousand times" (Adams, 1969, p. 329). Recently, the National Heart, Lung, and Blood Institute sponsored a conference on relapse prevention designed to focus research attention on this topic (Shumaker & Grunberg, 1986). Largely as a result of this conference research activities are now addressing this problem from a variety of bio-behavioral perspectives. If improved relapse-prevention strategies come out of these research activities, they should be incorporated into work-site smoking-cessation programs.

SUMMARY AND CONCLUSIONS

This chapter has reviewed the current state of affairs regarding smoking cessation at work sites. Smoking is clearly a serious health risk to smokers, and interaction with other exposures (e.g., asbestos) at some work sites increases this health risk. Despite the common belief that involuntary smoking is dangerous, currently there is little hard evidence that environmental tobacco smoke at most work sites poses serious health risks to most nonsmokers, especially when ventilation is good.

Employers who provide smoking-cessation programs certainly are doing a great service for employees who smoke and are helping to clean up indoor environments from toxins that may disturb nonsmoking employees. Smoking-cessation programs are more than an altruistic act by employers because the costs of smoking (e.g., absenteeism, health insurance) can be reduced if smoking rates decrease. Unfortunately, work-site smoking-cessation programs can be a mixed blessing. While some employees applaud the program, others may resent it as a fringe benefit or time-off work for others. To avoid these problems and the costs of cessation programs, employers may decide to ban or restrict smoking to particular places at work without providing a smoking-cessation program. This approach, however, does not consider the addictive nature of tobacco use and potential deleterious effects of cessation. In general, addictive behaviors cannot be effectively stopped solely by legislative or regulatory bans. Treatment programs are important.

There is a variety of smoking-cessation programs currently being offered at different work sites. Because few of these programs have been evaluated in controlled intervention trials, it is impossible to know which type of program is most effective. It seems that incentive and competition programs increase participation and are reasonably effective at work sites. Work-site programs probably could be improved by teaching participants to deal with consequences of smoking cessation that contribute to smoking relapse. Also, designers of work-site programs should attend to current research on smoking that may yield improved cessation strategies. In addition, designers of work-site smoking-cessation and other health promoting programs should draw from other fields (such as industrial and organizational psychology) to motivate employees to participate in the programs.

As a final note, I would like to add a more personal observation. It strikes me that American culture today is dominated by two, potentially conflicting, zeitgeists. One is a concern about health and fitness and the realization that our health is largely the result of our own behaviors. The second is a focus on financial security and material wealth. The 1980s will be remembered for well-groomed young investment bankers jogging and eating overpriced low-cholesterol foods at lunchtime. If my generalization is basically correct, then employers will provide smoking-cessation and other health-promoting programs as long as they are unquestionably cost-effective and rewarding to the business. There is, however, another reason to provide the most effective programs: They save lives and decrease suffering from major illnesses.

REFERENCES

Abrams, D. B., Pinto, R. P., Monti, P. M., Jacobus, S., Brown, R., & Elder, J. P. (1985). *Health education vs. social network support for relapse prevention in a worksite smoking cessation program.* Poster presented at the annual meeting of the Society of Behavioral Medicine, New Orleans, LA.

Adams, A. K. (Ed.). (1969). *The home book of humorous quotations.* New York: Dodd Mead.

Benowitz, N. L. (1983). The use of biologic fluid samples in assessing tobacco smoke consumption. In J. Grabowski & C. S. Bell (Eds.), *Measurement in the analysis and treatment of smoking behavior (NIDA Research Monograph 48*; pp. 6–26). Washington, DC: U.S. Government Printing Office.

Bowen, D. J., Spring, B., & Fox, E. (1989). Dietary advice and smoking cessation. Manuscript submitted for publication.

Bishop, D. B., & Fisher, E. B. (1984). *Employer assisted smoking elimination: The second year (9/83–10/84).* Annual report to the American Lung Association of Eastern Missouri, St. Louis.

Dawley, H. H., Jr., Fleischer, B. J., & Dawley, L. T. (1984). Smoking cessation with hospital employees: An example of worksite smoking cessation. *International Journal of the Addictions, 19,* 327–334.

Ellis, B. H. (1980). Prerequisites for successful workplace-based smoking cessation programs. *Proceedings of Ohio Department of Health Conference "Smoking and the Workplace,"* Columbus, OH.

Fernstrom, J. D., & Wurtman, R. J. (1971). Brain serotonin content: Increase following ingestion of carbohydrate diet. *Science, 174,* 1023–1025.

Fernstrom, J. D., & Wurtman, R. J. (1972). Brain serotonin content: Physiological regulation by plasma neutral amino acids. *Science, 178,* 414–416.

Fielding, J. E. (1982). Effectiveness of employee health improvement programs. *Journal of Occupational Medicine, 24,* 907–916.

Fielding, J. E. (1984). Health promotion and disease prevention at the worksite. *Annual Review of Public Health, 5,* 237–265.

Garfinkel, L., Auerbach, O., & Joubert, L. (1985). Involuntary smoking and lung cancer: A case-control study. *Journal of the National Cancer Institute, 75,* 463–469.

Glasgow, R. E., Klesges, R. C., Godding, P. R., Vasey, M. W., & O'Neill, H. K. (1984). Evaluation of a worksite controlled smoking program. *Journal of Consulting and Clinical Psychology, 52,* 137–138.

Glasgow, R. E., Klesges, R. C., & O'Neill, H. K. (in press). Programming social support for smoking modification: An extension and replication. *Addictive Behaviors.*

Grove, D. A., Reed, R. W., & Miller, L. C. (1979). A health promotion program in a corporate setting. *Journal of Family Practice, 9,* 83–88.

Grunberg, N. E. (1986). Nicotine as a psychoactive drug: Appetite regulation. *Psychopharmacology Bulletin, 22,* 875–881.

Grunberg, N. E. (1990). The inverse relationship between tobacco use and body weight. In L. T. Kozlowski, H. M. Annis, H. D. Cappell, F. B. Glaser, M. S. Goodstadt, Y. Israel, H. Kalant, E. M. Sellers, & E. R. Vingilis (Eds.). *Research Advances in Alcohol and Drug Problems* (pp.273–315). New York: Plenum Press.

Grunberg, N. E., & Bowen, D. J. (1985). Coping with the sequelae of smoking cessation. *Journal of Cardiopulmonary Rehabilitation, 5,* 285–289.

Grunberg, N. E., Bowen, D. J., & Winders, S. E. (1986). Effects of nicotine on body weight and food consumption in female rats. *Psychopharmacology, 90,* 101–105.

Grunberg, N. E., Winders, S. E., & Popp, K. A. (1987). Sex differences in nicotine's effects on consummatory behavior and body weight in rats. *Psychopharmacology, 91,* 221–225.

Heckler, L. M. (1980). Employee education programs—one aspect of nurse's expanded role in an occupational health program. *Occupational Health Nursing, 28,* 25–29.

Jason, L. A., Gruder, C. L., Martino, S., Flay, B., Warnecke, R., & Thomas, N. (1987). Work site group meetings and the effectiveness of a televised smoking cessation intervention. *American Journal of Community Psychology, 15*(1), 57.

Kabat, G. C., & Wynder, E. L. (1984). Lung cancer in nonsmokers. *Cancer, 53,* 1214–1221.

Kanzler, M., Zeidenberg, P., & Jaffe, J. H. (1976). Response of medical personnel to an on-site smoking cessation program. *Journal of Clinical Psychology, 32,* 670–674.

Klesges, R. C., Cigrang, J., & Glasgow, R. E. (1987). Worksite smoking modification programs: A state-of-the-art review and directions for future research. *Current Psychology Research & Reviews, 6,* 26–56.

Klesges, R. C., & Glasgow, R. E. (1986). Smoking modification in the worksite. In M. F. Cataldo & T. J. Coates (Eds.), *Health and industry: A behavioral medicine perspective* (pp. 231–254). New York: Wiley Interscience.

Klesges, R. C., Glasgow, R. E., Klesges, L. M., Morray, K., & Quale, R. (1987). Competition and relapse prevention training in worksite smoking modification. *Health Education Research: Theory and Practice, 2,* 5–14.

Klesges, R C., Vasey, M. W., & Glasgow, R. E. (1985). *Evaluation of a worksite smoking competition program.* Citation paper presented at the annual meeting of the Society of Behavioral Medicine, New Orleans, LA.

Klesges, R. C., Vasey, M. W., & Glasgow, R. E. (1986). A worksite smoking modification competition: Potential for public health impact. *American Journal of Public Health, 76,* 198–200.

Koo, L. C., Ho, J. H. C., & Saw, D. (1984). Is passive smoking an added risk factor for lung cancer in Chinese women? *Journal of Experimental and Clinical Cancer Research, 3,* 277–283.

Kornitzer, M., Dramiaix, M., Kittel, F., & DeBacker, G. (1980). The Belgian Heart Disease Prevention Project: Changes in smoking habits after two years of intervention. *Preventive Medicine, 9,* 496–503.

Lee, P. N., Chamberlain, J., & Alderson, M. R. (1986). Relationship of passive smoking to risk of lung cancer and other smoke-associated diseases. *British Journal of Cancer, 54*(1), 97–105.

Li, V. C., Kim. Y. J., Ewart, C. K., Terry, P. B., Cuthie, J. C. Wood, J., Emmett, E. A., & Permutt, S. (1984). Effects of physician counseling on the smoking behavior of asbestos-exposed workers. *Preventive Medicine, 13,* 462–476.

Long, M. A. D., & Simone, S. S. (1985). *A comparison of hypnosis vs. relapse prevention in a worksite multicomponent smoking treatment.* Poster presented at the annual meeting of the Association for the Advancement of Behavior Therapy, Houston, TX.

Malott, J. M., Glasgow, R. E., O'Neill, H. K., & Klesges, R. C. (1984). Coworker social support in a worksite smoking control program. *Journal of Applied Behavior Analysis, 17,* 485–496.

Meyer, A. J., & Henderson, J. B. (1974). Multiple risk factor reduction in the prevention of cardiovascular disease. *Preventive Medicine, 3,* 225–236.

Nepps, M. M. (1984). A minimal contact smoking cessation program at the worksite. *Addictive Behaviors, 9,* 291–294.

Norton-Ford, J. D., & Schmitz, E. G. (1982). *Health promotion at the worksite: A quasi-controlled trial of smoking cessation.* Paper presented at the annual meeting of the Association for the Advancement of Behavior Therapy, Los Angeles, CA.

Rand, C., Stitzer, M., Bigelow, G., & Mead, A. (1984). *Contingent reinforcement for smoking abstinence.* Poster presented at the annual meeting of the American Psychological Association, Toronto, Ontario, Canada.

Rose, G., Heller, R. F., Pedoe, H. T., & Christie, D. G. S. (1980). Heart Disease Prevention Project: A randomized controlled trial in industry. *British Medical Journal, 280,* 747–751.

Rosen, G. M., & Lichtenstein, E. (1977). An employee incentive program to reduce cigarette smoking. *Journal of Consulting Psychology, 45,* 957–959.

Schlegel, R. P., Manske, S. R., & Shannon, M. E. (1983). *Butt out: Evaluation of the Canadian Armed Forces Smoking Cessation Program.* Paper presented at the Fifth World Conference on Smoking and Health, Winnipeg, Manitoba, Canada.

Scott, R. R., Denier, C. A., & Prue, D. M. (1983). *Worksite smoking intervention with health professionals.* Paper presented at the Association for the Advancement of Behavior Therapy Annual Convention, Washington, DC.

Shepard, D. S. (1980). *Incentives for not smoking—experience at the Speedcall Corporation: Preliminary report.* Boston: Harvard School of Public Health, Center for the Analysis of Health Problems.

Shumaker, S. A., & Grunberg, N. E. (1986). Proceedings of the National Working Conference on Smoking Relapse. *Health Psychology, 5* Supplement, 1–99.

Sorenson, G., & Pechacek, T. F. (1987). Attitudes toward smoking cessation among men and women. *Journal of Behavioral Medicine, 10,* 129–137.

Sorman, K. (1979). This quit-smoking program works. *American Lung Association Bulletin, 65,* 2–6.

Stachnik, T., & Stoffelmayr, B. (1983). Worksite smoking cessation programs: A potential for national impact. *American Journal of Public Health, 73,* 1395–1396.

Terry, L. T. (1971). The future of an illusion. *American Journal of Public Health, 61,* 233–240.

U.S. Department of Health, Education and Welfare. (1979). *Smoking and health: A report of the Surgeon General* (DHEW Publication No. PHS–79–50066). Washington, DC: U.S. Government Printing Office.

U.S. Department of Health and Human Services. (1982). *The health consequences of smoking: Cancer. A report of the Surgeon General* (DHHS Publication No. PHS–82–50179). Washington, DC: U.S. Government Printing Office.

U.S. Department of Health and Human Services. (1983). *The health consequences of smoking: Cardiovascular disease. A report of the Surgeon General* (DHHS Publication No. PHS–84–50204). Washington, DC: U.S. Government Printing Office.

U.S. Department of Health and Human Services. (1985). *The health consequences of smoking: Cancer and chronic lung disease in the workplace. A report of the Surgeon General* (DHHS Publication No. 85–50207). Rockville, MD: U.S. Government Printing Office.

U.S. Department of Health and Human Services. (1986). *The health consequences of involuntary smoking. A report of the Surgeon General* (DHHS Publication No. CDC–87–8398). Washington, DC: U.S. Government Printing Office.

U.S. Department of Health and Human Services. (1988). *The health consequences of smoking: Nicotine addiction. A report of the Surgeon General.* Washington, DC: U.S. Government Printing Office.

U.S. Department of Health and Human Services. (1989). *Reducing the health consequences of smoking: 25 years of progress. A report of the Surgeon General.* (DHHS Publication No. CDC–89–8411). Washington, DC: U.S. Government Printing Office.

U.S. Public Health Service. (1964). *Smoking and health. Report of the Advisory Committee to the Surgeon General of the Public Health Service* (PHS Publication No. 1103). Washington, DC: U.S. Government Printing Office.

U.S. Public Health Service. (1968). *The health consequences of smoking. A Public Service review: 1967* (PHS Publication No. 1696 Revised). Washington, DC: U.S. Government Printing Office.

Wu, A. H., Henderson, B. E., Pike, M. C., & Yu, M. C. (1985). Smoking and other risk factors for lung cancer in women. *Journal of the National Cancer Institute, 74,* 747–751.

Wurtman, R. J. (1987). Dietary treatments that affect brain neurotransmitters. *Annals of the New York Academy of Sciences, 499,* 179–190.

Wurtman, J. J., & Wurtman, R. J. (1979). Drugs that enhance central serotoninergic transmission diminish elective carbohydrate consumption by rats. *Life Sciences, 24,* 895–904.

7 Behavioral Treatment of Obesity at the Work Site

John P. Foreyt
Department of Medicine
Baylor College of Medicine

Georganna Leavesley
Department of Psychology
University of Houston

If behavioral principles are effective in the treatment and maintenance of weight loss, as some researchers suggest, then why is obesity not decreasing in prevalence in our society? In 1960–1962, the percentage of overweight persons 25–74 years of age was 27.4; in 1976–1980, the percentage had risen to 28.4 (U.S. Department of Health and Human Services, 1986). Either obese individuals are not familiar with or are not using the principles to help control their weight, or the principles themselves are not very effective in the control of obesity. This chapter will explore the role of behavioral principles in the treatment of obesity and attempt to explain why they have had only limited effectiveness with weight problems and why their introduction at the work site has been a failure to date.

THE NATURE OF THE PROBLEM OF OBESITY

Despite enormous societal and health-related emphasis on being thin, obesity continues to be among today's most difficult medical and psychological problems (Foreyt, 1987). The problem affects a substantial portion of the population, with estimates of incidence ranging from 15% to 50% of adult Americans (Van Itallie, 1979).

Consequences of obesity are numerous and troublesome; they include increased risk for a number of health problems (Bray, 1985; Weiss, 1984). Additional consequences lie in the area of disturbing social and psychological problems (Brownell & Foreyt, 1985; Wadden & Stunkard, 1985). The majority of the field favors theories about the etiology of obesity which integrate genetics, physiology, psychology, and cultural factors (Brownell & Foreyt, 1985). Such a consensus is indicative of little conclusive evidence in the area of any specific explanation of the nature of obesity, and demonstrates the multitude of questions in the area that remain unanswered.

Without definitive theories for obesity, a number of treatments have been proposed and investigated, from fad dieting to jaw wiring to weight-loss pajamas, as well as comprehensive treatment packages, the most widely used of which is behavioral treatment

BEHAVIORAL TREATMENT FOR WEIGHT CONTROL

Behavioral treatment of obesity has been the subject of much excitement and study (Jeffery, Wing, & Stunkard, 1978; Kingsley & Wilson, 1977; Stuart, 1980); programs with this approach generally yield weight loss and behavior change that are encouraging but often poorly maintained (Stalonas, Perri, & Kerzner, 1984). Weight loss during an 8–12-week program averages approximately 11 pounds (or 1–2 pounds per week); this loss is fairly well maintained at one year. Regrettably, only 25% of participants have continued to lose weight after the formal sessions (Brownell, 1982).

Review articles conclude that the short-term effectiveness of behavioral treatment for weight loss is consistent and significant, the moderate weight loss achieved is often maintained after one year, and the attrition from such programs is much less than other approaches to treatment for obesity (Brownell & Wadden, 1986; Foreyt, 1987; Wilson & Brownell, 1980). Typically, behavior-therapy programs for weight loss include training in self-monitoring of weight and intake, and related self-reward. Participants are also taught stimulus control for situations in which one eats, and for promotion of behaviors that promote weight loss. This stimulus control is targeted at modifying the meaning of food and eating, as well as emotions associated with eating. Training in nutrition and a balanced diet, reduced in calories and fat, is also included (Foreyt et al., 1982; Jeffery, Gerber, Rosenthal, & Lindquist, 1983).

Many studies have focused on the correlations between the use of specific program techniques and in-program weight-loss success; others have addressed the use of techniques as correlated with weight change or maintenance after formal treatment (Brownell, 1984; Foreyt, Goodrick, & Gotto, 1981). To examine the role of the various areas of behavioral techniques, studies dealing with specific techniques will be reviewed.

Self-Monitoring

Among all of the factors involved in behavioral treatment for weight loss, self-monitoring has been identified as the one that seems to be indispensable

(Craighead, 1985). Self-monitoring of intake has been the factor most frequently found to predict weight loss or maintenance in follow-up studies
(Leon, 1977; Mahoney, 1974; Quereshi, 1977; Weiss, 1977). Although review
articles describe the importance of self-monitoring in weight loss, there is no
evidence that self-monitoring alone produces long-term weight loss and maintenance (Wilson & Brownell, 1980).

Self-monitoring measures for weight control have taken a variety of forms,
where various treatment aspects are monitored—specifically, program technique adherence on the whole, adherence to the prescribed diet, and adherence
to exercise routines. The monitoring can range from a mental monitoring of
weekly weights to daily written records of all intake and behaviors related to
eating (Leon, 1977). For example, in a study where the control group received
placebo attention and self-monitored by keeping food records (Harris & Hallbauer, 1973), there was significant weight loss in the control group at the close
of 12 weeks of treatment. Unfortunately, there was no maintenance of the
weight loss at seven-month follow-up.

Self-monitoring as a general approach to weight-loss maintenance was
found to be rated most helpful in a one-year follow-up of 108 behavioral program clients (Jeffery, Vender, & Wing, 1978). Weight loss in this program
correlated, at the close of treatment, with self-monitoring and controlling situations for eating, but there were no correlates found for weight-loss maintenance at the one-year follow-up. Similarly, in another study of behavioral
treatment subjects, weight loss at program close was correlated to self-
monitoring (Bellack, Schwartz & Rozensky, 1974). In Wing and Jeffery's
study of successful losers and maintainers (1978), close monitoring of weight
was a technique used by 75% of the respondents, while 25% of the subjects
used recording of intake.

A related technique involves self-reinforcement, in that rewarding oneself
for behavior requires that the behavior be monitored in some way. In a review
of studies predicting successful weight reduction, Weiss (1977) found self-
reinforcement style to hold the most promise as an indicator of potential for
responsiveness to a weight-control program. Persons termed high self-
reinforcers tended to have better results. Such a reinforcement style may be
indicative of a tendency to monitor oneself more closely.

Stalonas, Johnson, and Christ (1978) utilized self-reinforcement in the contingency portion of their treatment designs and found those subjects in groups
with this treatment factor to maintain loss at one year. Further, the group with
exercise plus the self-reinforcement (and inferred self-monitoring) was the
most successful at weight loss and one-year maintenance.

In a comparison of simple self-management techniques (counting and recording the number of bites of food taken daily) and combined self-
management techniques (weight and food monitoring, stimulus control, self-
reinforcement, and peer involvement), the groups did not differ from

each other in program success, but both exceeded a nonspecific treatment group and a no-treatment control group in effects at the end of treatment and at three-month follow-up. However, at the six-month follow-up, there were no group differences and all had regained (Hall, Hall, Hanson, & Borden, 1974).

Several studies have examined monitoring of weight as a regular behavior among weight-loss maintainers. For male weight-loss maintainers, daily weighing was reported as a strategy utilized by 85%, while 66% of the female respondents used this style of self-monitoring (Colvin & Olson, 1983). In other follow-up studies, continuing to keep weight records was a significant correlate of weight maintenance (Gotestam, 1979; Stalonas et al., 1984). Other studies identified weight monitoring as serving as an "early warning system" for weight-loss maintenance, alerting the person to institute measures to address any weight gain thusly noted (Stuart & Guire, 1978). Additionally, the use of an awareness of weight as a signal to restart dieting was reported as a maintenance technique by a significant number of the weight-loss maintainers in Marston and Criss's study (1984).

Investigating the effect of "nonspecific" factors in behavioral weight-control programs (Kirschenbaum, Stalonas, Zastowny, & Tomarken, 1985), researchers separated 65 subjects into four groups: receiving nonspecific treatment, prompts for attention to eating outside of therapy sessions, standard behavior therapy, and behavior therapy plus induced positive expectancies. At the close of treatment and three months later, there was no difference in weight loss among the four groups.

As the preceding discussion has illustrated, self-monitoring, despite being the most touted method for weight control, remains poorly understood. The problem of persons reporting less accurately in a "one-shot" follow-up interview is involved in many of these studies, and some fail to follow up after a long enough interval, as demonstrated by the repeated call in the literature for improvement in follow-up studies after behavioral programs (Colvin & Olson, 1984; Stalonas et al., 1984; Stalonas & Kirschenbaum, 1985).

Stimulus Control

The principle of stimulus control is the second component of behavioral treatment for weight loss to be considered. Stimulus-control behaviors include storing food out of sight, avoiding other activities while eating, eating only in one place, using a list to grocery shop, not shopping while hungry, leaving food on the plate, eating slowly, and using a small plate. The usefulness of these activities is uncertain. Some studies have looked at the effect of these various stimulus-control factors and found little evidence of their effectiveness. For example, in Wing and Jeffery's study (1978) of successful weight-loss maintainers, techniques for stimulus control were rarely utilized.

Considering stimulus control along with contingency management for weight loss, Stalonas et al. (1978) found weight loss and maintenance success in groups of behavior treatment participants that used stimulus control and self-reinforcement for utilizing the techniques of the program. This correlation was found at the close of formal classes and at 3 and 12 months after. In a five-year follow-up of this same study (Stalonas et al., 1984), only the "pure-experience" eating (avoiding other activities, eating only in one place) was correlated to weight loss. Once again, the results concerning this behavioral treatment component, stimulus control, are inconclusive and contradictory.

Contingency Management

Contracts and other forms of self-reward are used in behavioral treatment of obesity to effect contingencies related to weight-loss behavior. The Stalonas et al. study (1984) is one example of contingency management research.

Group and individual contracts for weight loss were researched with 89 male subjects who participated in a behavioral treatment for weight loss (Jeffery et al., 1983). The most successful strategy was group contracts; persons in the groups using contracts had losses maintained at one year. Unfortunately, other studies of contracts were less successful in maintenance.

Harris and Bruner (1971) found a greater weight loss in contract participants over self-controlled subjects during the formal program, although at ten months there was no difference between the groups. Correspondent with these findings, Colvin, Zopf, and Myers (1983) examined 23 co-workers in two groups, one with behavior therapy and monetary contracts and another with only the behavior therapy. They found that the contracted group had more in-program success, but the difference again faded at six-month follow-up. There are, once more, few conclusions that can be drawn on the basis of empirical evidence.

Dieting

Proceeding with the components of the typical behavioral treatment for weight control, dieting has played a central role in weight-loss programs for quite some time. Nutrient combinations have been touted for achieving balances that will reduce fat, enhance metabolism, or increase thermogenesis (Weiss, 1984). The diet most generally supported by research is well-balanced nutritionally and reduced in fat and calorie content as compared to the average American diet (Quereshi, 1977). Although weight-control techniques have progressed beyond mere counting calories, dieting continues to be the basis for most weight-control programs (Van Itallie, 1980).

Several studies have investigated dietary adherence as related to weight loss and weight-loss maintenance. Massively obese persons were studied 12 to

18 months after an extensive behavioral treatment program. Correlates of weight loss and maintenance included average caloric intake—those with less intake had better success at maintenance (Katahn, Pleas, Thackery, & Wallston, 1982).

Wing and Jeffery (1978) studied 64 successful subjects who had lost 20 or more pounds and maintained the loss more than a year. Losses were reported as maintained by monitoring weight and modifying food intake.

While these findings argue for the investigation of the effect of consistent effort at dieting as a factor in maintenance of weight change, it is noteworthy that the vast majority of studies do not address the role of diet. Nevertheless, it is assumed to be the cornerstone of the effective weight-control package, with a constellation of techniques targeted to enhance compliance with the diet.

Exercise

Exercise has been promoted as a treatment for obesity for its role in energy expenditure, metabolic rate adjustment (during and after exercise), and thermogenesis, as well as contributing to a loss of fat rather than lean body mass (Thompson, Jarvie, Lakey, & Cureton, 1982). Additionally, exercise may reduce some of the complications that may result from obesity (Brownell & Stunkard, 1980) and may also serve to enhance self-concept and stress management (Stern & Lowney, 1986). Unfortunately, chronic drawbacks to exercise programs for weight management are poor adherence and general attrition (Brownell & Stunkard, 1980; Stern & Lowney, 1986).

Although some studies have related exercise and increased activity to weight loss and maintenance (Perri, McAdoo, McAllister, Lauer, & Yancey, 1986; Stern & Lowney, 1986), many more studies have focused on behavioral techniques and failed to attend to the effects of exercise (Thompson et al., 1982). However, newer designs of behavioral therapy for weight loss tend to utilize a component of exercise (Foreyt, 1987).

In three studies, exercise or commitment to a more active lifestyle was correlated with weight-loss maintenance (Graham, Taylor, Howell, & Seigel, 1983; Katahn et al., 1982; Stuart & Guire, 1978). Another study looked at 48 participants in a behavioral weight-loss program who had been grouped and received the standard program, or combinations of the standard program with exercise and self-reinforcement (Stalonas et al., 1978). All groups lost significant amounts of weight during the formal treatment and had maintained this loss at the three-month follow-up. However, at one year after treatment, only the groups with exercise and/or self-reinforcement were successful at maintenance. The effect of exercise at one year was noticeable in that the group with exercise alone was the most successful at mean weight loss at each interval, end of program, 3 months, and 12 months after.

A similar study considered exercise and behavior therapy and post-treatment contact in a crossed design in which groups received behavioral treatment for weight loss with or without aerobic exercise and did or did not receive post-treatment contact (Perri et al., 1986). Among the 90 subjects, those in the groups with aerobic exercise lost more weight in the program and showed better weight-loss progress after treatment as well. This was found in spite of the trend to decrease in self-reported adherence to the exercise program in the post-treatment period of 18 months.

Two studies of weight-loss subjects, some of whom were independent losers and some program losers, have found correlations between exercise and maintenance of weight. Marston and Criss (1984) studied 47 formerly overweight persons who predominantly reported exercising several times weekly, among their maintenance strategies. Similarly, Colvin and Olson (1983) interviewed persons who had lost 20% or more of their total weight and maintained the loss for two years. Eighty-five percent of the male respondents cited vigorous exercise as a factor in weight maintenance, compared with 78% of the female respondents.

Additionally, a study of 40 persons, 7 months after behavioral treatment for weight loss, found persons who used frequent exercise were successful at maintaining weight loss (Gormally, Rardin, & Black, 1980). Interestingly, in-program weight loss was not correlated with an increase in activity level. Also grouping subjects (into remediably and irremediably obese categories), Quereshi (1977) found a more active lifestyle to be a characteristic of the remediably obese group.

One study looked at 46 weight-loss program volunteers and investigated the difference between groups; one group used a behavior therapy program with an exercise component, one without exercise, and one was an attention/placebo control group (Harris & Hallbauer, 1973). All three groups lost weight during the formal programming. The groups receiving treatment lost more at seven months than did the control group, and the group that utilized exercise lost an average of five pounds more than the treatment group without exercise.

A somewhat different, but related study was conducted by Epstein, Wing, Koeske, and Valoski (1984), in which 53 families with children were involved in one of three groups: one with diet, one with diet and exercise, and one no-treatment control group. Parents showed a greater in-program weight loss in the exercise and diet group. At a 12-month follow-up, it was found that exposure to the exercise treatment was predictive of weight-loss maintenance.

Epstein and colleagues (Epstein, Wing, Koeske, Ossip, & Beck, 1982) have shown that individuals are more likely to add exercise behaviors that are part of their current lifestyles, such as increasing the amount of time they walk, rather than adding some vigorous aerobic component like jogging to their lives. Although clearly an important part of any long-term behavioral change program for weight maintenance, most obese individuals do not add a regular exercise component to their lifestyles.

The many benefits of exercise are intuitively apparent, including improved cardiovascular health, stress management, compliance with current social mores for physical fitness, and other personal satisfactions. It has not been made clear what constitutes effective exercise, leading to little knowledge as to what type, duration, intensity, and frequency of exercise to promote for weight control.

TRENDS IN THE BEHAVIORAL TREATMENT OF OBESITY

Increasing Weight Losses

Today, the major trend in the behavioral treatment of mild obesity (i.e., less than 50 pounds overweight) is increasing the length of the formal treatment. It is not uncommon to see programs of 24 weeks in length, consisting of weekly group classes. Since behavioral programs result in average losses of about 1.2 pounds per week, losses of 25 to 30 pounds at post treatment are not unusual. These losses are quite encouraging and, if maintained, suggest a real breakthrough in the treatment of mild obesity.

Unfortunately, the two major problems in this area are dropouts during treatment, resulting in data being reported only from individuals who complete the program (usually those who are losing weight), and, of course, maintenance of the losses. Given the large number of persons who receive behavioral treatment for their obesity, the number of long-term follow-up studies reported is still quite small. However, of those available, it appears that individuals who complete treatment gain back about one-third of their losses one year later. Follow-up data are much rarer beyond one year.

For individuals who need to lose more than 50 pounds, the major trend has been the very-low-calorie diets (VLCDs) which consist of 200–800 kcal per day, primarily protein, along with electrolyte, mineral and vitamin supplementation (Blackburn, Lynch, & Wong, 1986). These VLCDs either consist of protein in a liquid form, using a milk- or egg-based formula, or protein from low-fat meat, chicken, turkey, or seafood. Both the liquid-formula and meat approach result in equivalent very rapid weight losses. Individuals lose about 45 pounds after 12 weeks in either approach. When individuals are monitored closely by physicians, the VLCDs appear reasonably safe for up to 12 weeks (Wadden, Stunkard, & Brownell, 1983). Although many individuals stay on their VLCDs for periods exceeding 12 weeks, some up to a year and more, studies assessing safety factors are lacking.

The major problem with the VLCDs has been their inability to show maintenance of the lost weight following the reintroduction of a balanced diet. Published studies report rapid regain in short periods of time. There have been a few encouraging exceptions. Wadden and Stunkard (1986), for example, in a controlled study, reported that individuals who received only a VLCD lost 31 pounds, those who received only behavior therapy 31.5 pounds, and those who received both VLCD and behavior therapy, 42.5 pounds. Most importantly, the individuals in the two behavioral conditions were better able to maintain their losses, although all of the groups showed some regain. Such studies are unfortunately rare; most work in this area shows rapid regain, even after receiving behavioral training.

Weight-Loss Maintenance

As has been demonstrated repeatedly in this discussion of the literature, maintenance of weight loss is a persistent problem. Results of some studies have addressed weight-loss maintenance as related to the degree of post-treatment use of behavioral techniques.

In a 15-month follow-up of 721 Weight Watchers who had reached goal weight, weight-loss maintenance was correlated with a reported greater general use of the self-monitoring and stimulus-control techniques taught in the formal program (Stuart & Guire, 1978). However, these results should be generalized with caution, in that subjects who achieve goal weight are atypical to the majority of behavioral treatment subjects.

Another study, with a much smaller sample, studied participants of a behavioral weight-control program at 1 and 3 years after the program (Gotestam, 1979). Of the 11 subjects, only 1 affirmed using the program's behavioral techniques during the post-program period, and 7 stated that they had "thought about" the techniques. The differences were small; those who used techniques had a mean weight loss of about four pounds at 3 years, and those who did not had a mean weight gain of almost four pounds.

In a similar study, with a sample of 60 being followed 4.5 years after treatment, weight-loss maintenance was correlated with reported adherence to at least three behavioral procedures (Graham et al., 1983). Fifty-eight percent of the subjects reported using techniques; those who were physically active lost almost five pounds during the follow up period and were the best group at maintaining their in-program losses. Again, it is not clear which techniques were used as the authors speak of adherence scores rather than specific techniques.

In a very thorough study, a 5-year follow-up of behavioral treatment was conducted with 36 subjects contacted to assess correlates of weight-loss maintenance (Stalonas et al., 1984). Although the amount (6 out of 11–13) and consistency (24% of potential use) of behavioral techniques utilized was low, those subjects who reported using techniques had better success at weight-loss maintenance. Specifically, self-monitoring and stimulus-control measures were used most often.

A different type of study was conducted by Sjoberg and Persson (1979). Nine participants in a weight-loss program (using dieting, education, and support) were interviewed, during the 4 months after the start of the program, about techniques they had used to lose weight. Those using a greater number of techniques were found to have had better success at weight loss. However, the format of the treatment used in this study precludes drawing conclusions to behavioral treatment or any other program where techniques are taught and thereafter utilized.

The problem of weight-loss maintenance has received much attention in the form of research on strategies for continuing the effects of formal programming into the maintenance period. With the theory that longer treatment would result in better learning and long-term use of the behavior-modifying techniques, Jeffery, Wing, and Stunkard (1978) extended the length of their behavioral programs to 29 weeks. The extended treatment group showed more in-program loss; but the long-term success showed no advantage over the 10-week program.

A number of researchers have considered ''booster sessions'' as a method to enhance long-term treatment effects, with inconsistent results. Reviewing studies, two had favorable results and five failed to show lasting weight losses (Wilson, 1980). Bimonthly booster sessions, which included group support and self-monitoring with feedback from the therapists, were compared with self-monitoring and mailing the results to the therapists, and with minimal contact for weights at intervals after treatment (Hall, Bass, & Monroe, 1978). Although the booster group showed significant differences at 2, 4, and 6 months, the difference was not present at the one-year follow-up.

Another strategy utilized to enhance maintenance is reliance on social-support systems (Mahoney & Mahoney, 1976; Wilson & Brownell, 1978). The results of such studies have been mixed. A number of studies have suggested that the addition of spouses or other family members or close friends in the intervention may improve the individual's ability to lose weight and maintain the losses (Brownell, Heckerman, Westlake, Hayes, & Monti, 1978; Pearce, LeBow, & Orchard, 1981; Rosenthal, Allen, & Winter, 1980; Saccone & Israel, 1978). Not all studies have supported the addition of a social-support individual in weight-loss programs (Brownell & Stunkard, 1981; O'Neill et al., 1979, Weisz & Bucher, 1980). The complexity of defining support makes it difficult to appropriately assess its role with obese persons. Simply insisting that the spouse attend weight-loss classes may or may not be perceived as supportive by an obese individual. Building and maintaining support systems is frequently beyond the scope of many short-term behaviorally oriented weight-loss intervention programs.

Relapse-prevention training (Marlatt & Gordon, 1985) aims at teaching people how to deal with high-risk situations which tend to make it difficult for them to stay within their behavior-control program. Stressful situations and unfamiliar settings, for example, frequently lead individuals astray. Relapse-prevention training in the treatment of obesity has been a topic of recent focus (Brownell, Marlatt, Lichenstein, & Wilson, 1986; Craighead, 1985). A number of studies which have included relapse-prevention training in their weight-loss program have shown some evidence of improved long-term maintenance (Abrams & Follick, 1983; Perri, McAdoo, Spevak, & Newlin, 1984; Perri, Shapiro, Ludwig, Twentyman, & McAdoo, 1984). Other studies have not supported this effectiveness (Collins, Rothblum, & Wilson, 1986).

In a study of maintenance strategies (Perri et al., 1984b), 129 subjects were assigned to six different approaches. Treatment approaches included behavior therapy; behavior therapy combined with relapse-prevention training; and a treatment of diet, exercise, and nonspecific treatment effects. One group from each approach received post-treatment therapist contact. The nonspecific treatment group and the group receiving behavior therapy plus relapse-prevention training showed enhanced weight loss at 1 year when they had received post-treatment contact. Conversely, the group receiving behavior therapy alone did not show any difference as a result of contact.

In the study by Perri et al. (1986), post-treatment conditions were researched. Those subjects not exposed to the multi-component maintenance program reported a decrease in adherence to the program at 3 months, while the group that had the maintenance program only showed a decrease at the 12-month follow-up. The authors propose that the maintenance contact program may delay relapse, rather than prevent it.

SUMMARY OF BEHAVIORAL TREATMENT FOR OBESITY

Unanswered questions abound in the field of treatment for weight control: which techniques to use, what their effect is, which behaviors to promote, which to discourage, what are desirable short-term expectations and what can be hoped for in terms of long term maintenance, what factors play a role in individual differences . . . the list seems endless. A great number of studies have endeavored to reduce the list of questions, with little success. The diversity of persons attempting to lose weight, as well as the complex nature of the problem contributes to the paucity of conclusive research results.

Some recommendations can be drawn from the literature, however. Longer interventions result in larger weight losses. A six-month treatment program consisting of 24 weekly group sessions should result in an average weight loss of 25 to 30 pounds. It appears that one longer program of this length results in better weight losses than several programs of shorter duration.

More comprehensive programs seem to keep individuals in the program for longer periods of time. Today's intervention programs tend to emphasize nutrition education, exercise behaviors, behavior modification, social support, and cognitive-behavioral change strategies (Foreyt, 1987).

Emphasizing maintenance strategies appears to reduce the rebound effect following the end of formal intervention. Potential maintenance strategies include training in relapse prevention, building long-term social-support systems, and adding exercise to one's daily lifestyle.

For individuals more than 50 pounds overweight, the use of very-low-calorie diets appears to be promising for increasing the number of pounds lost in reasonably short periods of time, while motivation is still quite high. Unfortunately, maintenance of the lost weight has been poor. Behavioral strategies have not been particularly successful to date at preventing the inevitable rebound in most individuals treated.

APPLICATION TO THE WORK SITE

Intervention at the work site seems to make good sense. Employees are a captive audience, and making the program highly accessible should ensure a good turnout. Unfortunately, early work-site programs for obese employees showed just the opposite; attrition was very high and weight losses very modest.

Abrams and Follick (1983) reported 48% attrition along with an average weight loss of only 7 pounds after a 10-week program; Brownell, Stunkard, and McKeon (1985) did not fare much better with 42% attrition and an average weight loss of 7.9 pounds after 16 weeks of intervention. Remarkably similar data were reported by Fisher, Lowe, Levenkron, and Newman (1982), that is, 40% attrition and an average weight loss of 9.9 pounds after a 10-week program; and by Follick, Fowler, and Brown (1984), 40% attrition and an average weight loss of 9.7 pounds after a 14-week intervention. Something was very wrong. High numbers of dropouts and relatively low weight losses were consistently occurring.

These similarities in data indicate that some component of the program must have been responsible. Clinical programs involving the same elements and identical lengths were showing attrition rates around 11% (Brownell & Wadden, 1986) and weight losses of about 1.2 pounds per week. These clinical programs were developed for highly motivated persons who, oftentimes with great difficulty, made the trek to a university clinic or, more frequently, went to a commercial program and paid large amounts of money for treatment. Apparently, these programs were not suitable for this new, different clientele, individuals who may not have been particularly motivated to either attend classes with regularity or do the large amounts of homework behavioral programs require.

What are the differences between the work site and the clinic? One of the characteristics of the work site is the opportunity for competition with colleagues one knows quite well and sees daily (Stunkard, 1986). A number of clever studies by Brownell and colleagues (Brownell, Cohen, Stunkard, Felix, Cooley, 1984) suggested that competition among employees results in very low attrition and good weight losses. For example, they reported the results of a work-site competition held among three banks. Weight-loss goals were purposely modest, a five-dollar entrance fee was charged, and a large bulletin board in each bank's lobby displayed weekly progress. The prize for the winning bank was the entrance fee of all participants. Results showed a remarkably low dropout rate of only 0.5% and an average weight loss of 13 pounds after the 12-week program. These competitions are quite inexpensive to run and seem to result in weight losses comparable to clinic programs of the same length.

The major problem of weight-loss competitions at the work site is the lack of maintenance. Competition programs do not ordinarily teach behavioral skills needed to keep the weight off following the competition. Competitions tend to reward weight loss, but do not necessarily teach individuals how to most effectively achieve that loss or, more importantly, how to maintain the loss following the completion of the competition. Clearly, behavioral training is needed following the competition; how to recruit and maintain work-site employees in such programs has yet to be solved. Until the work-site programs begin to show reasonable maintenance following intervention, they must be regarded as a failure, similar to earlier programs relying only on reinforcement strategies (see, for example, Foreyt, 1977).

CONCLUSIONS

Clinically, the addition of behavioral principles to weight-loss programs for the treatment of mild obesity has improved results and reduced attrition. Their usefulness with moderate and severe obesity, however, is less clear. The initial application of these behavioral principles to the work site in the form of clinical programs developed at university clinics resulted in high dropout rates and relatively low weight losses. The use of competition between work sites has shown some positive results, with attrition being remarkably low and weight losses consistent with clinic programs of similar lengths. Unfortunately, unless individuals also receive training in behavioral principles, regain of lost body weight following the completion of the competition appears to be inevitable. To date, programs aimed at the long-term control of obesity at the work site have been a failure. One of the greatest challenges still facing behavioral scientists today is the development of effective strategies for helping individuals to conquer this complex, refractory disorder.

ACKNOWLEDGMENT

Preparation of this chapter was supported in part by Grant 1 RO1 HL 33954–01 from the National Heart, Lung, and Blood Institute of the National Institutes of Health.

REFERENCES

Abrams, D. B., & Follick, M. J. (1983). Behavioral weight loss intervention at the work site: Feasibility and maintenance. *Journal of Consulting and Clinical Psychology, 51,* 226–233.

Bellack, A. S., Schwartz, J., & Rosensky, R. H. (1974). The contribution of external control to self-control in a weight reduction program. *Journal of Behavior Therapy and Experimental Psychiatry, 5,* 245–249.

Blackburn, G. L., Lynch, M. E., & Wong, S. L. (1986). The very-low-calorie diet: A weight-reduction technique. In K. D. Brownell & J. P. Foreyt (Eds.), *Handbook of eating disorders: Physiology, psychology, and treatment of obesity, anorexia, and bulimia* (pp.198–212). New York: Basic Books.

Bray, G. A. (1985). Complications of obesity. *Annals of Internal Medicine, 103,* 1052–1062.

Brownell, K. D (1982). Obesity: understanding and treating a serious, prevalent, and refractory disorder. *Journal of Consulting and Clinical Psychology, 50,* 820–824.

Brownell, K. D. (1984). The addictive disorders. *Annual Review of Behavior Therapy: Theory and Practice, 8,* 208–272.

Brownell, K. D., Cohen, R. Y., Stunkard, A. J., Felix, M. R., & Cooley, N. B. (1984). Weight loss competitions at the work site: Impact on weight, morale, and cost-effectiveness. *American Journal of Public Health, 74,* 1283–1285.

Brownell, K. D., and Foreyt, J. P. (1985). Obesity. In D. H. Barlow (Ed.), *Clinical handbook of psychological disorders* (pp. 299–343). New York: Guilford.

Brownell, K. D., Heckerman, C. L., Westlake, R. J., Hayes, S. C., & Monti, P. M. (1978). The effect of couples training and partner cooperativeness in the behavioral treatment of obesity. *Behaviour Research and Therapy, 16,* 323–333.

Brownell, K. D., Marlatt, G. A., Lichtenstein, E., & Wilson, G. T. (1986). Understanding and preventing relapse. *American Psychologist, 7,* 765–782.

Brownell, K. D., & Stunkard, A. J. (1980). Physical activity in the development of and control of obesity. In A. J. Stunkard (Ed.), *Obesity* (pp. 300–324). Philadelphia: Saunders.

Brownell, K. D., & Stunkard, A. J. (1981). Couples training, pharmaco-therapy, and behavior therapy in the treatment of obesity. *Archives of General Psychiatry, 38* 1224–1229.

Brownell, K. D., Stunkard, A. J., & McKeon, P. E. (1985). Weight reduction at the work site: A promise partially fulfilled. *American Journal of Psychiatry, 141,* 47–51.

Brownell, K. D., & Wadden, T. A. (1986). Behavior therapy for obesity: Modern approaches and better results. In K. D. Brownell & J. P. Foreyt (Eds.), *Handbook of eating disorders: Physiology, psychology, and treatment of obesity, anorexia, and bulimia* (pp. 180–197). New York: Basic Books.

Collins, R. L., Rothblum, E., & Wilson, G. T. (1986). The comparative efficacy of cognitive and behavioral approaches to the treatment of obesity. *Cognitive Therapy and Research, 10,* 299–318.

Colvin, R. H., & Olson, S. B. (1983). A descriptive analysis of men and women who have lost significant weight and are highly successful at maintaining the loss. *Addictive Behaviors, 8,* 287–295.

Colvin, R. H., & Olson, S. B. (1984). Winners revisited: An 18 month follow up of our successful weight losers. *Addictive Behaviors, 9,* 305–306.

Colvin, R. H., Zopf, K. J., & Myers, J. H. (1983). Weight control among workers. *Behavior Modification, 7,* 64–75.

Craighead, L. W. (1985). A problem-solving approach to the treatment of obesity. In M. Hersen & A. S. Bellack (Eds.), *Handbook of clinical behavior therapy with adults* (pp. 229–268). New York: Plenum.

Epstein, L. H., Wing, R. R., Koeske, R., Ossip, D., & Beck, S. (1982). A comparison of lifestyle change and programmed aerobic exercise on weight and fitness changes in obese children. *Behavior Therapy, 13, 651–665.*

Epstein, L. H., Wing, R. R., Koeske, R., & Valoski, A. (1984). Effects of diet plus exercise on weight change in parents and children. *Journal of Consulting and Clinical Psychology, 52,* 429–437.

Fisher, E. D., Jr., Lowe, M. R., Levenkron, J. C., & Newman, A. (1982). Reinforcement and structural support of maintained risk reduction. In R. B. Stuart (Ed.), *Adherence, compliance and generalization in behavioral medicine,* (pp. 145–168). New York: Brunner/Mazel.

Follick, M. J., Fowler, J. L., & Brown, R. (1984). Attrition in work site interventions: the effects of an incentive procedure. *Journal of Consulting and Clinical Psychology, 52,* 139–140.

Foreyt, J. P. (1977). Therapist reinforcement techniques: Introduction. In J. P. Foreyt (Ed.), *Behavioral treatments of obesity* (pp.231–234). New York: Pergamon Press.

Foreyt, J. P. (1987). Issues in the assessment and treatment of obesity. *Journal of Consulting and Clinical Psychology, 55,* 677–684.

Foreyt, J. P., Goodrick, G. K., & Gotto, A. M. (1981). Limitations of behavioral treatment of obesity: Review and analysis. *Journal of Behavioral Medicine, 4,* 159–174.

Foreyt, J. P., Mitchell, R. E., Garner, D. T., Gee, M., Scott, L. W., & Gotto, A. M. (1982). Behavioral treatment of obesity: Results and limitations. *Behavior Therapy, 13,* 153–161.

Gormally, J., Rardin, D., & Black, S. (1980). Correlates of successful response to a behavioral weight control clinic. *Journal of Counseling Psychology, 27,* 179–191.

Gotestam, K. G. (1979). A three year follow up of a behavioral treatment for obesity. *Addictive Behaviors, 4,* 179–183.

Graham, L. E., Taylor, C. B., Howell, M. F., & Seigel, W. (1983). Five year follow up to a behavioral weight loss program. *Journal of Consulting and Clinical Psychology, 46,* 368–369.

Hall, S. M., Bass, A., & Monroe, J. (1978). Continued contact and monitoring as follow-up strategies: A long-term study of obesity treatment. *Addictive Behaviors, 3,* 139–147.

Hall, S. M., Hall, R. G., Hanson, R. W., & Borden, B. L. (1974). Permanence of two self-managed treatments of overweight in university and community populations. *Journal of Consulting and Clinical Psychology, 42,* 781–786.

Harris, M B., & Bruner, C. G. (1971). A comparison of a self-control and a contract procedure for weight control. *Behaviour Research and Therapy, 9,* 347–354.

Harris, M. B., & Hallbauer, E. S. (1973). Self-directed weight control through eating and exercise. *Behaviour Research and Therapy, 11,* 523–529.

Jeffery, R. W., Gerber, W. M., Rosenthal, B. S., & Lindquist, R. A. (1983). Monetary contracts in weight control: Effectiveness of group and individual contracts of varying size. *Journal of Consulting and Clinical Psychology, 51,* 242–248.

Jeffery, R. W., Vender, M., & Wing, R. R. (1978). Weight loss and behavior change 1 year after behavioral treatment for obesity. *Journal of Consulting and Clinical Psychology, 46,* 368–369.

Jeffery, R. W., Wing, R. R., & Stunkard, A. J. (1978). Behavioral treatment of obesity: The state of the art, 1976. *Behavior Therapy, 12,* 144–149.

Katahn, M., Pleas, J., Thackery, M., & Wallston, K. A. (1982). Relationship of eating and activity self-reports to follow-up weight maintenance in the massively obese. *Behavior Therapy, 13,* 521–528.

Kingsley, R. G., & Wilson, G. T. (1977). Behavior therapy for obesity: A comparative investigation of long-term efficacy. *Journal of Consulting and Clinical Psychology, 45,* 288–298.

Kirschenbaum, D. S., Stalonas, P. M., Zastowny, T. R., & Tomarken, A. J. (1985). Behavioral treatment of adult obesity: Attentional controls and a 2-year follow-up. *Behaviour Research and Therapy, 23,* 675–682.

Leon, G. R. (1977). A behavioral approach to obesity. *The American Journal of Clinical Nutrition, 30,* 785–789.

Mahoney, M. J. (1974). Self-reward and self-monitoring techniques for weight control. *Behavior Therapy, 5,* 48–57.

Mahoney, M. J., & Mahoney, K. (1976). Treatment of obesity: A clinical exploration. In B. J. Williams, S. Martin, & J. P. Foreyt (Eds.), *Obesity: Behavioral approaches to dietary management* (pp. 30–39). New York: Brunner/Mazel.

Marlatt, G. A., & Gordon, J. R. (Eds.). (1985). *Relapse prevention: Maintenance strategies in the treatment of addictive behaviors.* New York: Guilford.

Marston, A. R., & Criss, J. (1984). Maintenance of successful weight loss: Incidence and prediction. *International Journal of Obesity, 8,* 435–439.

O'Neill, P. M., Currey, H. S., Hirsch, A. A., Riddle, F. E., Taylor, C. I., Malcolm, R. J., & Sexauer, J. D. (1979). Effects of sex of subject and spouse involvement on weight loss in a behavioral treatment program: A retrospective investigation. *Addictive Behaviors, 4,* 167–178.

Pearce, J. W., LeBow, M. D., & Orchard, J. (1981). Role of spouse involvement in the behavioral treatment of overweight women. *Journal of Consulting and Clinical Psychology, 49,* 236–244.

Perri, M. G., McAdoo, W. G., McAllister, D. A., Lauer, J. B., & Yancey, D. Z. (1986). Enhancing the efficacy of behavior therapy for obesity: Effects of aerobic exercise and a multicomponent maintenance program. *Journal of Consulting and Clinical Psychology, 54,* 670–675.

Perri, M. G., McAdoo, W. G., Spevak, P. A., & Newlin, D. B. (1984a). Effects of a multicomponent maintenance program on long-term weight loss. *Journal of Consulting and Clinical Psychology, 52,* 480–481.

Perri, M. G., Shapiro, R. M., Ludwig, W. W., Twentyman, S. T., & McAdoo, W. G. (1984b). Maintenance strategies for the treatment of obesity: An evaluation of relapse training and posttreatment contact by mail and telephone. *Journal of Consulting and Clinical Psychology, 52,* 404–413.

Quereshi, M Y. (1977). Psychosocial correlates of obesity control. *Journal of Clinical Psychology, 33,* 343–350.

Rosenthal, B., Allen, G. J., & Winter, C. (1980). Husband involvement in the behavioral treatment of overweight women: Initial effects and long-term follow-up. *International Journal of Obesity, 4,* 165–173.

Saccone, A. J., & Israel, A. C. (1978). Effects of experimenter versus significant other controlled reinforcement and choice of target behavior on weight loss. *Behavior Therapy, 9,* 271–278.

Sjoberg, L., & Persson, L. (1979). A study of attempts by obese patients to regulate eating. *Addictive Behaviors, 4,* 349–359.

Stalonas, P. M., Johnson, W. G., & Christ, M. (1978). Behavior modification for obesity: The evaluation of exercise, contingency management, and program adherence. *Journal of Consulting and Clinical Psychology, 46,* 463–469.

Stalonas, P. M., & Kirschenbaum, D. S. (1985). Behavioral treatment for obesity and eating habits revisited. *Behavior Therapy, 16,* 1–14.

Stalonas, P. M., Perri, M. G., & Kerzner, A. B. (1984). Do behavioral treatments of obesity last? A five-year follow-up investigation. *Addictive Behaviors, 9,* 175–184.

Stern, J. S., & Lowney, P. (1986). Obesity: The role of physical activity. In K. D. Brownell & J. P. Foreyt (Eds.), *Handbook of eating disorders: Physiology, psychology, and treatment of obesity, anorexia, and bulimia* (pp. 145–158). New York: Basic Books.

Stuart, R. B. (1980). Weight loss and beyond: Are they taking it off and keeping it off? In P. O. Davidson & S. M. Davidson (Eds.), *Behavioral medicine: Changing health lifestyles* (pp. 151–194). New York: Brunner/Mazel.

Stuart, R. B., & Guire, K. (1978). Some correlates of the maintenance of weight lost through behavior modification. *International Journal of Obesity, 2,* 225–235.

Stunkard, A. J. (1986). The control of obesity: Social and community perspectives. In K. D. Brownell & J. P. Foreyt (Eds.), *Handbook of eating disorders: Physiology, psychology, and treatment of obesity, anorexia, and bulimia* (pp. 213–228). New York: Basic Books.

Thompson, J. K., Jarvie, G. J., Lakey, B. B., & Cureton, K. J. (1982). Exercise and obesity: Etiology, physiology, and intervention. *Psychological Bulletin, 91,* 55–75.

U.S. Department of Health and Human Services. (1986). *Health: United States 1985* [DHHS Publication No. (PHS) 86–1232]. Washington, DC.: U.S. Government Printing Office.

Van Itallie, T. B. (1979). Obesity: Adverse effects on health and longevity. *American Journal of Clinical Nutrition, 32,* 2723–2733.

Van Itallie, T. B. (1980). Dietary approaches to the treatment of obesity. In A. J. Stunkard (Ed.), *Obesity* (pp. 249–261). Philadelphia: Saunders.

Wadden, T. A., & Stunkard, A. J. (1985). Social and psychological consequences of obesity. *Annals of Internal Medicine, 103,* 1062–1067.

Wadden, T. A., & Stunkard, A. J. (1986). Controlled trial of very low calorie diet, behavior therapy, and their combination in the treatment of obesity. *Journal of Consulting and Clinical Psychology, 54,* 482–488.

Wadden, T. A., Stunkard, A. J., & Brownell, K. D. (1983). Very low calorie diets: Their efficacy, safety and future. *Annals of Internal Medicine, 99,* 675–684.

Weiss, A. R. (1977). Characteristics of successful weight reducers: A brief review of predictor variables. *Addictive Behaviors, 2,* 193–201.

Weiss, S. R. (1984). Obesity. *Psychiatric Clinics of North America, 7,* 307–319.

Weisz, G., & Bucher, B. (1980). Involving husbands in the treatment of obesity: Effects on weight loss, depression, and marital satisfaction. *Behavior Therapy, 11,* 643–650.

Wilson, G. T. (1980). Behavior modification and the treatment of obesity. In A. J. Stunkard (Ed.), *Obesity* (pp.325–344). Philadelphia: Saunders.

Wilson, G. T., & Brownell, K. D. (1978). Behavior therapy for obesity: Including family members in the treatment process. *Behavior Therapy, 9,* 943–945.

Wilson, G. T., & Brownell, K. D. (1980). Behavior therapy for obesity: An evaluation of treatment outcome. *Advances in Behavioral Research and Therapy, 3,* 49–86.

Wing, R. R., & Jeffery, R. W. (1978). Successful losers: A descriptive analysis of the process of weight reduction. *Obesity/Bariatric Medicine, 7,* 190–191.

8 Exercise Programs at Work: Data, Issues, and Models

William L. Haskell
Stanford Center for Research in Disease Prevention
Stanford University School of Medicine

INTRODUCTION

Over the past several decades, the idea that exercise should be included as one component of a comprehensive program of health promotion has become generally accepted. It appears that this acceptance is due, at least in part, to the increase in scientific evidence that moderate amounts of exercise contribute to improved health status and, despite the "fitness boom" of the past decade, that there is still a large segment of the adult population which is sedentary or which exercises only sporadically. Recent research has also continued to define the type and amount of exercise that is of benefit, and investigators have begun to identify personal and program features that enhance the adoption of health-oriented exercise and its long-term maintainence. Some of this research and program development has been conducted at the work site with quite favorable preliminary results being reported not only for improvement in functional capacity and health status, but also for reductions in health-care costs and absenteeism.

This chapter contains an overview of the major health benefits of exercise, including some of the key issues yet to be resolved and a brief review of recent reports directed at the results of selected work-site exercise programs. Given the experience gained to date, some of the issues regarding work-site exercise program implementation are presented and several program models are discussed.

FITNESS REQUIREMENTS FOR OCCUPATIONAL WORK

To successfully perform most jobs in the United States requires very little in the way of human physical effort. Technological advances in manufacturing, construction, transportation, farming, mining, and many other industries have substantially reduced the number of workers that are required to perform vigorous activity in order to make a living. For the many workers performing white-collar jobs, the most vigorous job-related exercise they are required to perform is getting to work; and for many blue-collar workers, machines, computers, or robots reduce their jobs to pushing buttons or pulling levers while sitting or standing. The peak energy requirement for most of these jobs does not exceed 3 METS (multiples of resting energy expenditure) or 4 kilocalories per minute, which is less than 35% of the aerobic capacity of most healthy adults. For these individuals, physical working capacity is not directly a factor in their job performance, and increasing endurance capacity or strength by exercise training is very unlikely to increase their productivity. For workers required to perform heavy physical labor, such as lifting and carrying heavy objects, shoveling or digging for extended periods, or using heavy tools, to be most effective in these jobs they need an above-average level of physical fitness. Maintenence of physical fitness for these jobs is especially important for older workers, due to the general age-related reduction in physical working capacity. For workers who perform vigorous tasks on a daily basis, the work itself is probably a sufficient training stimulus for maintaining their working capacity, and an ancillary physical conditioning program is not necessary. For workers who are required to perform vigorous activity less frequently, such as firefighters or police, it would seem that for optimal job performance, a conditioning program to improve and maintain endurance and strength is required.

Workers who have a substantial reduction in physical working capacity due to chronic degenerative disorders, such as ischemic heart disease, chronic obstructive lung disease, arthritis, or adult-onset diabetes, may be able to perform a wider variety of jobs if their capacity is enhanced by physical training. Rehabilitative exercise can reduce exertion-induced symptoms, like shortness of breath, muscle fatigue, and chest pain, making the work experience much more acceptable and probably safer. Recent studies have demonstrated that an aggressive approach to physical rehabilitation can significantly accelerate return to full employment after acute myocardial infarction, and the cost savings of this approach are substantial (Dennis, Houston-Miller, Schwartz, et al, 1988).

The primary objective of including a physical activity component in a work-site health promotion program is not to directly increase worker productivity by increasing their physical fitness. For a majority of employees, the goal is for the exercise to contribute to the prevention of the clinical manifestations of various chronic degenerative disorders or to improve their general psychological status in ways that would lead to better employee attitude or morale and enhanced productivity. Many short-term biologic and psychologic effects of exercise have been reasonably well established (Haskell, 1985; Phelps, 1987), but what has been much more difficult is the documentation that these benefits lead to increased worker productivity, reduced absenteeism, or decreased health-care costs.

EXERCISE AND HEALTH

The fact that people who are more physically active on the job or during leisure time possess better health than their less active counterparts continues to be demonstrated by various lines of research. Prospective observational studies comparing the morbidity and mortality status of more versus less active persons find lower rates of chronic degenerative disease in the more active, and this relationship appears to be independent of other known risk factors. Recent intervention studies in which sedentary adults are randomized to exercise training or remaining sedentary have observed a wide variety of exercise-induced beneficial effects, either directly on physical or psychological health or on biologic functions that very likely lead to an improved clinical status. Health problems for which scientific data exist, demonstrating that exercise provides some protective effect or is useful in their treatment, include coronary heart disease, hypertension, adult-onset diabetes, osteoporosis, obesity, and psychological dysfunction, including depression and anxiety. Other areas of great interest but for which only preliminary data exist and which require much more research before considering a possible causal link to inactivity include cancer (Garabrant, Peters, Mack, & Bernstein, 1984) and a reduced immune response (Simon, 1984).

Coronary Heart Disease

Evidence continues to mount in support of the hypothesis that participation in moderately intense physical activity provides some protection against the clinical manifestations of coronary heart disease (CHD). A recent comprehensive review of the scientific literature by Powell and colleagues (Powell, Thompson, Caspersen, & Kendrick, 1987) concluded:

> the inverse association between physical activity and incidence of CHD is consistently observed, especially in the better designed studies: This association is appropiately sequenced, biologically graded, plausible, and coherent with existing knowledge. . . . These observations suggest that in CHD prevention pro-

grams, regular physical activity should be promoted as vigorously as blood pressure control, dietary modification to lower serum cholesterol, and smoking cessation. Given the large proportion of sedentary persons in the United States, the incidence of CHD attributable to insufficient physical activity is likely to be surprisingly large. (p. 283)

The results of several major studies since the publication of this review also support these conclusions.

An important contribution to the exercise and CHD hypothesis was the recent report by Leon and colleagues (Leon, Connett, Jacobs, & Rauramaa, 1987) on 12,138 men enrolled in the Multiple Risk Factor Intervention Trial and followed for an average of seven years. The results of this report are summarized in Table 8.1. Leisure-time physical activity (LTPA) was quantitated

TABLE 8.1
Leisure-Time Physical Activity and Coronary Heart Disease (CHD): The Multiple Risk Factor Intervention Trial

	Physical Activity Tertiles		
	1	2	3
CHD Deaths*	24.6	15.4	15.4
CHD Risk Ratio	1.0	0.63	0.65
Physical Activity			
Calories/day	74	224	638
Minutes/day	15	47	134
Light (minutes/day)	5	18	53
Moderate (minutes/day)	6	17	41
Heavy (minutes/day)	3	9	28
Other (minutes/day)	1	3	12

Source: Leon, Connant, Jacobs, and Rauramaa (1987).
*Age-adjusted rate per 1,000; $N = 12,318$

over the year preceding entry into the study and expressed as mean minutes of physical activity per day at light, moderate, or high intensity. When total LTPA was divided into tertiles, those men in the middle tertile (moderately active) had 63% as many fatal CHD events and sudden deaths and 70% as many deaths as men in the lowest tertile of LTPA ($p < .01$). Mortality rates with high LTPA were similar to those in moderate LTPA, but combined fatal and nonfatal major CHD events were 20% lower with high as compared to low LTPA ($p < .05$). These associations were not substantially weakened when adjusted for other major CHD risk factors. Similar results for CHD and total mortality have been reported for Harvard University alumni in several followup reports by Paffenbarger and colleagues (Paffenbarger, Hyde, Wing, & Hsieh, 1986).

Short-term exercise training studies continue to provide information on how exercise might alter various biologic functions that would lead to reduced

CHD risk. Of greatest interest, based on the number of reports, has been the modification of the plasma lipoprotein profile by endurance exercise training. The major changes with exercise appear to be an increase in high-density lipoprotein cholesterol concentration and the associated apolipoprotein A-1 and a decrease in triglyceride concentration when initially elevated. What mediates these changes has not yet been fully established, but the enzymes lipoprotein lipase and lecithin cholesterol acetyltransferase appear to play an important role. Other mechanisms of interest include the effect of exercise on platlet aggregation and blood viscosity, enhanced insulin sensitivity, decreased catecholamine sensitivity leading to a decrease in myocardial workload and electrical instability, and reduced blood pressure. A summary of the major mechanisms being proposed for how exercise reduces the clinical manifestations of CHD has been published (Haskell, 1985).

Hypertension

Men and women who exhibit higher levels of aerobic capacity (Blair, Goodyear, Gibbons, & Cooper, 1984b) or are more physically active (Paffenbarger, Wing, Hyde, & Hsieh, 1983) exhibit lower systemic arterial blood pressure than less-fit or inactive persons of a similar age. Studies observing this association usually have demonstrated that it is somewhat confounded by the more-active or fit individuals also having a lower body mass index or percentage of body fat. Numerous exercise training studies have investigated the effects of endurance or strength training on resting blood pressure in normotensive men with an almost unanimous negative result: No significant reduction is observed if the study has an adequate non-exercising control group and the exercising subjects do not lose weight (Tipton, 1984). In contrast to the data on normotensive persons, a majority of studies that have investigated the effects of endurance exercise training in patients with essential hypertension have observed a reduction in both diastolic and systolic blood pressure (Seals & Hagberg, 1984). The average reduction in blood pressure in the hypertensive patients was approximately 10 mm Hg systolic and 8 mm Hg diastolic. However, as with many of the training studies on normotensive individuals, many of the studies on hypertensive patients have major design flaws that reduce the confidence that can be placed in their results.

Several recent training studies have been well designed and support the notion that exercise training can lower blood pressure in hypertensive patients, independent of changes in dietary intake or body weight (Duncan, et al., 1985; Nelson, Esler, Jennings, & Korner, 1986). Duncan et al. (1985) investigated the effects of a 16-week endurance exercise program on the blood pressure of patients with mild hypertension classified as either normoadrenergic or hyperadrenegic. Changes in systolic blood pressure were 6.3, 10.3, and 15.5 mm Hg for non-exercising controls, normoadrenegic and hyperadrenegic patients, respectively. In the hyperadrenegic group, decreases in total plasma catecholamine concentration with training were significantly associated with decreases in systolic ($r = .51$ $p < .05$) and diastolic ($r = .62$; $p < .01$) blood pressure. These data suggest that the exercise training reduced the endogenous release of norepinephrine in the hyperadrenergic patients. Nelson et al. (1986) also observed a significant decrease in both systolic and diastolic blood pressure after only four weeks of exercise training at either three or seven times per week (45-minute sessions at 60–75% of working capacity or cycle ergometer). Body weight and 24-hour sodium excretion remained constant, while plasma norepinephrine concentrations and peripheral vascular resistance fell in response to both training regimens.

It appears that moderate-intensity exercise training, less than that frequently recommended to increase maximal oxygen uptake, may be adequate to lower blood pressure in hypertensive patients. Roman, Camuzzi, Villalon, and Klenner (1981) observed that the blood pressure of hypertensive women was lowered to the same extent by 12 months of either low- or high-intensity training. Also, Kiyonaga, Arakawa, Tanaka, and Shindo (1985) reported significant reductions in systolic and diastolic pressures in essential hypertensive patients following a mild to moderate exercise training program. These and other similar data (Seals & Hagberg, 1984) indicate that in somewhat older hypertensive patients, activities such as brisk walking might confer a significant blood-pressure-lowering effect. If this is the case, it would be an important consideration in the design of appropiate exercise regimens.

Glucose Intolerance/Diabetes

Following the second decade of life, there tends to be a decrease in glucose tolerance, with the plasma glucose concentration two hours after an oral glucose tolerance test increasing approximately 5 mg/100 ml each decade. This decrease in glucose tolerance can be associated with an increase in plasma insulin concentrations or hyperinsulinemia. This increase in insulin may be due to both an increase in insulin production and decreased clearance. Allowed to continue for some years, this abnormal glucose/insulin response may lead to adult-onset or non-insulin-dependent diabetes. It now appears that diet, obesity, and exercise all can significantly affect the insulin response to glucose challenge by modifying insulin-mediated glucose removal. Whether regular exercise can help prevent adult-onset diabetes has not been established.

Endurance exercise training results in lower plasma insulin concentrations both while fasting and for several hours after glucose or food consumption (Heath, et al., 1983). This decreased insulin response to a glucose challenge in highly trained athletes or after a moderate to vigorous exercise training program indicates that tissue sensitivity and/or responsiveness to insulin is improved (LeBlanc, Nadeau, Richard, & Tremblay, 1981). This response appears to be due both to a decrease in insulin secretion by the pancreas and a more rapid glucose uptake at the same insulin concentration by the tissues, especially the muscles. This increase in insulin responsiveness appears to be more of an acute or transient effect of exercise than a chronic training effect. Most of this improvement in insulin action will disappear within several days after the last bout of exercise (Heath, et al., 1983). This response would argue for the exercise component of a health-promotion program to be performed at least on an every-other-day basis. As compared to the moderate-intensity exercise that might be adequate to reduce blood pressure, preliminary data indicate that somewhat higher-intensity exercise may be needed to enhance insulin action to a clinically significant level.

Patients with either insulin-dependent or non-insulin-dependent diabetes can benefit in terms of their clinical management from regular exercise training. The non-insulin-dependent patients have an increased insulin action after training, whereas insulin-dependent patients frequently require less insulin and may reduce their overall coronary heart disease risk status (Kemmer & Berger, 1983). These patients need to have their exercise regimen intergrated into their overall medical management, and the initial exercise sessions should be conducted under medical supervision. Given the significant effect of exercise on insulin action, a change in exercise habits can significantly alter the treatment plan for a diabetic patient.

Bone Density/Osteoporosis

Bone strength decreases in older adults due to a loss of bone-mineral content, especially calcium. This loss of mineral in men usually begins at around age 50 and progresses quite slowly (approximately 0.4–0.5 % per year). In women, however, it may begin as early as age 35 and occur at the rate of about 1% per year until menopause. During the first 4 to 5 years immediately following menopause, this loss accelerates to 2–4 % per year and then slowly returns to the premenopausal rate. By age 70, some women may lose up to 70% of their bone-mineral mass. The clinical term for this bone loss is "osteoporosis," and it is the primary cause of the increased rate of bone fractures in older adults. The exact mechanism that produces this demineralization of the bone has not been established, but contributing factors include hormonal changes, nutritional factors, and mechanical forces consisting of gravity and action of muscles.

Exercise will not prevent all of the bone-mineral loss associated with aging, but there is increasing evidence that bone-mineral content varies significantly with changes in gravitational or muscular forces acting on the bone. Extended periods of weightlessness experienced during space flight produce significant bone-mineral loss (Mack, LeChance, Vose, & Vogt, 1967), as does extended bed rest (Donaldson, et al., 1970). The opposite effect has been shown as well: With vigorous exercise, the mineral content and cross-sectional area of bones in the exercising limb and the spinal column are substantially increased

(Jones, Priest, Hayes, Tichenor, & Nagel, 1977; Montoye, Smith, Fardon, & Howley, 1980).

Moderate-intensity exercise, well within the capacity of many older adults, appears to either increase or at least maintain total body calcium and bone-mineral content. When Smith and colleagues (Smith, Reddan, & Smith, 1981) had elderly women exercise 30 minutes per session, 3 days per week for 3 years, they observed an increase in bone-mineral content of the radius of 2.3%. Twelve women who remained sedentary during this time showed a decrease of 3.3% ($p < .005$). The results of other short- and long-term exercise training studies in middle-aged and older women suggest that exercise stimulates bone-mineral retention and should be included as part of a comprehensive program of osteoporosis prevention (Krolner, Tondevold, Toft, Berthelsen, & Nielsen, 1982; Smith, Smith, Ensign, & Shea, 1981). Since the time of most-rapid bone-mineral loss in women is during the first four to five years after menopause, this may be a critical time for exercise intervention. Since many women are employed at this time, work-site exercise programs may want to consider this need in their plans.

Obesity

Health professionals are beginning to develop a much better appreciation for the potential role of exercise in the successful management of obesity. The value of exercise becomes especially important in dealing with long-term weight control for optimal health and performance in a society that encourages people to be inactive as a result of automation and makes high-density caloric foods readily available at a relatively low cost. It is now more frequently recognized that within a highly industrialized culture, those individuals who consume the greatest number of calories tend to be the leanest, while the more overweight individuals consume fewer calories (Yano, Rhoads, Kagan, & Tilleison, 1978). This is especially true when the caloric intake of highly active adults, such as joggers or tennis players, is compared to their sedentary counterparts. The more active men and women are significantly leaner, yet consume up to 25% more calories per day (Blair, et al., 1981). Along with these calories come more essential nutrients, which reduces the likelihood of any nutritional defeciency, an important consideration for people with poor eating habits who are trying to stay thin.

Exercise performed on a frequent basis has the potential to influence body composition by several different mechanisms. First, the actual calories expended during even moderate-intensity activity if performed frequently can produce a significant caloric deficit. Walking briskly for 10 minutes three times per day five days per week results in energy expenditure of approximately 1,000 calories per week or 15 pounds of body fat per year. In addition, there is some evidence that, following exercise of moderate to vigorous intensity, the metabolic rate remains elevated for some hours, thus further increasing the contribution of exercise to weight loss.

It is unlikely that the resting or basal metabolic rate of a cell that does not change in size as a result of exercise is increased. Thus, if exercise training does not increase muscle mass, then basal metabolic rate probably is not increased. However, because exercise can help retain or increase muscle mass (more in men than in women and more with strength than with endurance-type exercise), the potential exists for exercise to increase basal metabolic rate. There are two situations in which this effect may be quite important: first, in retaining lean body mass as a person grows older; and second, combining exercise with caloric restriction during a period of weight loss helps retain lean body mass, causing a greater proportion of the weight loss to be fat tissue (Zuti & Golding, 1976).

It has been suggested by some investigators that exercise may reset the appetite-control center so that a better balance exists between caloric intake and need (Woo, Garrow, & Pi-Sunyer, 1982). When activity decreases to a certain level, appetite and caloric intake do not decrease proportionately, and adiposity increases. Very active people consume more calories than do sedentary individuals, yet the former have a more optimal body composition (Blair, et al., 1981). When individuals are sufficiently active, it is likely that they can respond to the social pressures to eat and consume the readily available high-caloric foods and still not achieve a state of positive caloric balance. Under these conditions, appetite once again may be "allowed" to regulate caloric intake. Also, preliminary evidence from our laboratory suggests that participation in an exercise program by overweight men and women increases their adherence to a weight-reduction diet.

Psychological Benefits

A variety of claims have been made regarding the psychological benefits of physical activity, ranging from its use in the treatment of major psychological disorders to a euphoric state produced by running (runner's high). Actually, there is still very little known about the nature of the psychological changes due to exercise in an employed population, the frequency of these changes, and the characteristics of the exercise that produces them. The most frequently cited benefits include a reduction in depression and anxiety and a general increase in self-efficacy for exercise. Several major reviews have been published that have discussed some of the limitations of the data currently available and that have considered the type of research needed to better understand the independent contribution of physical activity to mental health and its relationship to worker productivity (Taylor, Sallis, & Needle, 1985).

In attempting to determine what types of activity should be included in a work-site exercise program with one objective being to enhance psychological status, a number of questions still need to be answered. Are the effects due to biochemical changes from the activity itself, are they behavioral and due to the interaction between the exerciser and the exercising situation, or are they some combination of these factors? Do some of the beneficial psychological effects require a specific biologic alteration, such as a decrease in sympathetic nervous system activity? Are others dependent on a behavioral stimulus, such as physical separation from a stress-producing situation or interaction with an exercise leader? From the little that is known about these issues, it appears that multiple stimuli exist.

As Bahrke and Morgan (1978) have noted, exercise may be a useful coping mechanism for some people as "time-out" therapy. Leisure-time activity can be an effective way of physically and mentally separating oneself from stress-producing situations. The physical separation, pleasant surroundings, enthusiastic and attractive exercise leaders, and sympathetic co-exercisers may be all that are required to decrease anxiety, hostility, or depression. The actual biologic effects of exercise, and thus the specific characteristics of exercise required to produce benefit, may be of lesser consequence. If this is the case, then activity performed someplace other than at the work site may be more effective in reducing job-related stress than if it were performed on site.

WORK-SITE EXERCISE PROGRAMS

Work-site exercise programs have been started for a number of reasons; some due to the fitness interests of the CEO or other top executive, including the corporate medical director, others because it was thought to be the proper thing to do for employees, or the employees obtained a program as a result of negotiations. Most recent programs have been implemented with the objective that they would increase corporate earnings by increasing employee productivity and/or decreasing health-care costs. Increases in physical fitness and a favorable alteration in risk factors for ischemic heart disease have been achieved by numerous programs (Haskell & Blair, 1980; Iverson, Fielding, Crow, & Christenson, 1985). This is not surprising, given that many work-site exercise program regimens were patterned after programs conducted in a research setting that had been shown to be effective for these purposes. What has been much more difficult to determine is the impact of various work-site exercise programs on long-term health outcomes, occupational productivity, and corporate economics.

Very few recent work-site exercise programs have been implemented in a manner that allows for the specific effects of the exercise to be isolated from those of other components of health promotion. Thus, other than for measures of physical fitness, it has not been possible to determine the specific effects of the exercise program in comprehensive work-site health-promotion programs

TABLE 8.2
Potential Job-Related Benefits of Work-Site-Based Physical Activity Programs

Potential Benefits	Level of Data Support*
Increased physical working capacity	+++
Inhanced attitude toward work	+
Improved job performance/productivity	+
Decrease in absenteeism	++
Decrease in health-care costs	+

*+ = minimal data to support
 ++ = data support likely cause-effect relationship
 +++ = definitive evidence of a benefit

such as Johnson & Johnson's Live for Life (Blair, Piserchia, Wilbur, & Crowder, 1986). Whereas a comprehensive approach to health promotion is probbly much more effective than isolated programs, more research is needed on specific components like exercise to better understand their independent effect on work-related outcomes.

Besides the various health-related benefits of exercise training discussed in the previous section, there are a number of proposed benefits that may be considered of value to the employer, as well as the employee. Listed in Table 8.2 are the benefits frequently cited as possibly resulting from participation in an exercise program that may increase the work-related productivity of the employee. For some of these items, very little supporting data have been collected using scientifically sound methods, and such evidence is exceedingly difficult to collect because of the complexity of the factors involved. The issues of work productivity and employee health-care costs both fall into this category. Other items, such as physical working capacity and attitude toward work, are much easier to document.

Physical Working Capacity

An increase in aerobic capacity has been used as a criterion for indicating that the type, intensity, and amount of exercise being performed is adequate to provide meaningful performance and health-related benefits. More is known about the characteristics of the exercise required to produce a significant increase in aerobic capacity than about any other outcome. Nearly all previously sedentary persons participating in a program of endurance exercise training consisting of 30 minutes of moderate-intensity exercise (50–75% maximal oxygen uptake), at least 3 times per week will experience a significant increase in physical working capacity within 12 weeks. Increases of 10–20% have been demonstrated for a wide variety of programs, including walking, jogging, cycling, swimming, circuit training, and aerobic dance in both men and women from age 20 to 70 years (American College of Sports Medicine, 1986).

Various work-site programs have demonstrated significant increases in physical working capacity in employee participants of an expected magnitude, based on the results of prior dose-response training studies (Cady, Thomas, & Karwasky, 1985; Fogle & Verdesca, 1975). Of special interest is the report from the Live for Life program (Blair, et al., 1986). Over a period of two years, employees in two companies were encouraged to participate in a broad-based health-promotion program, including physical activity, while employees in two other companies acted as controls. Employees in both sets of companies were given a fitness examination at baseline and annually for two years. As compared to baseline, participants in the health-promotion program reported an increase in physical activity at each of the two follow-up examinations and had a significant increase in maximal oxygen uptake, estimated from a sub-maximal exercise test (8.4% and 10.5% increase at year 1 and 2 vs. 1.5% and 4.7% in the health-screening-only employees). These changes occurred in both genders of all ages and across all education and socioeconomic classifications. This is the first objective evidence that a public-health approach at the work-site can produce clinically meaningful increases in physical activity and physical fitness.

Better Attitudes Toward the Work Environment

As observed with many non-work-site exercise programs, participation in work-site based exercise programs is frequently associated with enhanced psychological status, including better attitudes toward work and the work environment. In the USPHS-NASA exercise program (Durbeck et al., 1972), employees with the highest attendance and improved fitness reported their normal work routines as being less boring and enjoyed their jobs more. Cox, Shephard, and Corey (1981) found similar improvements in work-related attitudes in employees participating in work-site exercise program for six months. However, they did not find any change in a multifactor "job description index." Other studies have reported similar responses, but these studies usually have had not control or comparison group or were a quasi-experimental design (Blair, et al., 1984a).

Absenteeism

It would seem that documentation of the effects of a work-site exercise program on absenteeism would be relatively easy, but solid data are still lacking. The preliminary data look quite favorable with exercise program participants showing some reduction in paid sick leave and increases in physical fitness being correlated with decreased absenteeism. In an early study of New York State Education Department employees, a reduction of 4.7 hours per year per employee for all program participants was observed over the year following a 15-week exercise program compared to the prior "control year" (Bjurst-rom & Alexiou, 1978). Somewhat similar data have been reported by Cox, et al. (1981) among Canadian Life Insurance employees, and a smaller and more questionable difference between program participants and control-group employees.

Baum, Bernacki, and Tsai (1986) evaluated the absenteeism and health-care costs of 517 randomly selected participants and nonparticipants in the Tenneco health and fitness program. The illness-related absenteeism was lower for fe-male participants, as compared to nonparticipants (47 vs. 69 hours per year, $p < .05$ for women; 25 vs. 30 hours, $p = 28$ for men); but the differences occurred at the initiation of the program, thus representing primarily employee selection bias and not program effects. A physical-fitness program was in-cluded as part of a comprehensive health-promotion program provided to em-ployees of a large metropolitan school district in Texas (Blair et al., 1986). Of 12,136 employees, 3,846 enrolled in the program, and data on absenteeism were collected over the initial year of the program and compared to the prior year's experience. Participants who completed the health-promotion program had an average of 1.25 days less absenteeism ($p < .0001$) than did nonpartic-ipants (ANCOVA by using age, gender, ethnic status, and previous year absen-teeism as covariates). This was a comprehensive health-promotion program, so the change in absenteeism should not be totally ascribed to the exercise component, even though increases in treadmill exercise time were significantly associated with decreases in absenteeism.

Employee Health-Care Costs

As compared to absenteeism, employee health-care costs are very difficult to accurately track and can be influenced by numerous factors other than the physical health or psychological status of the employees. Several studies look-ing at short-term health-care costs (1–2 years) have reported reduced costs in program participants versus nonparticipants, but none of these studies has used a randomized design, and the results more likely represent participant selection bias and not program effects (Baum et al., 1986; Shephard, Corey, Renzland, & Cox, 1982). In several reports, the magnitude of the increase in fitness of program participants has been correlated with decreases in health-care costs, but the relationship could very well be due to program participant selection bias (Bly, Jones, & Richardson, 1986). Longer-term studies of fire-fighters in Los Angeles County (Cady et al., 1985), Blue Cross/Blue Shield employees in Indiana (Gibbs, Mulvaney, Henes, & Reed, 1985), and Johnson and Johnson employees in New Jersey (Blair et al., 1986) tend to support these shorter studies by demonstrating that fitness levels and program participation rates are inversely correlated with health-care costs. These data are enticing and make it all the more important to conduct well-designed, long-term trials to answer this very important question.

PROGRAM DESIGN AND IMPLEMENTATION:
MEDICAL VERSUS PUBLIC-HEALTH MODEL

Several reviews have been published that discuss many of the issues involved in the successful implementation of work-site exercise programs (Haskell & Blair, 1980; Fielding, 1984; Washington Business Group on Health, 1986). Instead of reiterating this material, it would seem to be more useful to consider the potential benefits of a medical or clinical approach to work-site fitness programming versus a public-health approach. While there are no set definitions for either of these approaches, the medical model is based on individual employee participation with emphasis placed on design features that meet the specific needs of each individual employee (evaluation, prescription, supervision, adherence monitoring, revaluation), whereas the public-health model emphasizes the overall needs of the employee population. The public-health model usually gives more attention to education, enhancement of a variety of exercise opportunities, and altering the environment to increase exercise opportunities and to reinforce desired behaviors. The benefits of the medical model are usually evaluated by the fitness changes made by only those employees who enroll in the program, but the public-health model would be evaluated by its effects on the entire work force. The integrated, community-based, health-promotion model used in the Five-City Project is an example of the public-health approach in the community (Farquhar et al., 1985), and the Live for Life program at Johnson and Johnson is an example of a work-site public-health approach (Bly et al., 1986).

Significant health/performance benefits of exercise at the work site will result only if employees make long-term changes in exercise habits lasting for years, rather than weeks or months. It is a reasonably rare experience for this type of change to occur in a large percentage of employees if the only exercise opportunities which are taken advantage of are at the work site. It seems that great success would be achieved by promoting a wide variety of exercise opportunities, including at home leisure time (gardening, active games with family members, etc.), participation in recreational sports on the employee's own time, and an increase in walking and cycling as a means of commuting to work, shopping, etc. Also, it is unrealistic to expect most people to begin an exercise program and not experience one or more relapses. Thus, it is important to teach relapse prevention as part of the program and develop a system of counseling to assist those people returning to their program when ready. Such relapses frequently occur when job or home situations change, following an illness or injury, or when lack of progress toward desired goals is perceived by the participant. In the public-health model, a major component is likely to be providing assistance to employees in starting or getting back to their own exercise program once they are ready. Opportunities need to be provided that meet the schedules of the participants rather than the schedule of the program planners. Thus, a program with an open enrollment design that features a wide variety of exercise options is likely to meet with greater success than a program with specific enrollment dates and restricted exercise options.

Since it now appears that at least some of the health- and possibly work-related benefits of exercise can be achieved by the performance of light- to moderate-intensity exercise, work-site-based programs should consider how to take advantage of this information. Men and especially women over age 45, who are not already regular exercisers, express a greater interest in light- or moderate-intensity exercise, such as brisk walking, than in more vigorous exercise, such as jogging. This lower-intensity exercise option lends itself to the public-health approach in that due to the lower medical risk, less medical screening and supervision is needed. It is now generally accepted that moderate-intensity activity (less than 50% of aerobic capacity) can be promoted among a working population without a medical examination being required first, as long as the necessary precautions are made to those people at increased medical risk during exercise. Not having to require or conduct medical examinations, especially exercise testing, prior to starting an exercise program eliminates a substantial logistic and cost barrier for many employees and employers.

The long-term success of a work-site exercise program can be enhanced in the public-health model by modification of the total work environment and experience to support a more active lifestyle for all employees. Minor changes in facilities can be made to increase the opportunity for walking or use of stairs, stationary exercise equipment can be decentralized from a core exercise facility to local areas in large corporations, and walking routes can be developed near the work-site and maps distributed to the employees. Some companies have incorporated flex time in order to reduce overuse of exercise facilities at peak hours, while others have contracted with local exercise facilities to allow employees to use their services or equipment at a reduced cost or even no cost. Using this broad approach, all employees are considered participants in the program, some just more active at specific times than are others. Over the long term, this public-health approach is more likely to have cost-effective benefits for the employer as well as the employee.

REFERENCES

American College of Sports Medicine. (1986). *Guidelines for graded exercise testing and training.* Philadelphia: Lea and Febiger.

Bahrke, M. E., & Morgan, W. P. (1978). Anxiety reduction following exercise and medication. *Cognitive Therapy Research, 2,* 323–334.

Baum, W. B., Bernacki, E. J., & Tsai, S. P. (1986). A preliminary investigation: Effect of a corporate fitness program on absenteeism and health care costs. *Journal of Occupational Medicine, 28,* 18–22.

Bjurstrom, L. A., & Alexiou, N. G. (1978). A program of heart disease intervention for public employees. *Journal of Occupational Medicine, 20,* 521–531.

Blair, S. N., Collingwood, T. R., Reynolds, R., Smith, M., Hagan, R. D., & Sterling, C. L. (1984a). Health promotion for educators: Impact on health behavior, satisfaction, and general well-being. *American Journal of Public Health, 74,* 147–149.

Blair, S. N., Ellsworth, N., Haskell, W. L., Stern, M., Farquhar, J., Wood, P. D. (1981). Comparison of nutrient intake in middle-aged men and women runners and controls. *Medicine and Science in Sports and Exercise, 13,* 310–315.

Blair, S. N., Goodyear, N. M., Gibbons, L. W., & Cooper, K. H. (1984b). Physical fitness and incidence of hypertension in healthy and normotensive men and women. *Journal of the American Medical Association, 252,* 487–490.

Blair, S. N., Piserchia, P. V., Wilbur, C. S., & Crowder, J. H. (1986). A public health intervention model for work-site health promotiion. *Journal of the American Medical Association, 255,* 921–926.

Blair, S. N., Smith, M., Collingwood, T. R., Reynolds, R., Prentice, M. D., & Sterling, C. L. (1986). Health promotion for educators: Impact on absenteeism. *Preventive Medicine, 15,* 166–175.

Bly, J. L., Jones, R. C., & Richardson, J. E. (1986). Impact of work-site health promotion on health care costs and utilization. *Journal of the American Medical Association, 256,* 3235–3240.

Cady, L. D., Thomas, P. C., & Karwasky, R. J. (1985). Program for increasing health and physical fitness of fire fighters. *Journal of Occupational Medicine, 27,* 110–114.

Cox, M., Shephard, R. J., & Corey, P. (1981). Influence of an employee fitness programme upon fitness, productivity and absenteeism. *Ergonomics, 24,* 795–806.

Dennis, C., Houston-Miller, N., Schwartz, R. G., Ahn, D., Kraemer, H. C., Gossard, D., Juneau, M., Taylor, C. B., DeBusk, R. F. (1988). Early return to work after uncomplicated myocardial infarction. *Journal of the American Medical Association, 260,* 214–220.

Donaldson, C. L., Hulley, S. B., Vogel, J. M., Hattner, R. S., Bayers, J. H., & McMillan, D. E. (1970). Effect of prolonged bed rest on bone mineral. *Metabolism, 19,* 1071–1084.

Duncan, J. J., Farr, J. E., Upton, J., Hagan, R. D., Oglesby, M. E., & Blair, S. N. (1985). The effects of aerobic exercise on plasma catecholamines and blood pressure in patients with mild essential hypertension. *Journal of the American Medical Association, 254,* 2609–2613.

Durbeck, D. C. Heinzelmann, F., Schacter, J., Haskell, W. L., Payne, G. H., Moxley, R. T., Nemiroff, M., Limoncelli, D. D., Arnoldi, L. B., Fox, S. M. (1972). The National Aeronautics and Space Administration-U.S. Public Health Service health evaluation and enhancement program (1972). *American Journal of Cardiology, 30,* 784–790.

Farquhar, J. W., Fortmann, S., Maccoby, N., Haskell, W. L., Williams, P. T., Flora, J. A., Taylor, C. B., Brown, B. W., Solomon, D. S., & Hulley, S. B. (1985). The Stanford Five City project: Design and methods. *American Journal of Epidemiology, 122,* 323–334.

Fielding, J. E. (1984). Health promotion and disease prevention at the work-site. *American Review of Public Health, 5,* 237–265.

Fogle, R. K., & Verdesca, A. S. (1975). The cardiovascular conditioning effects of a supervised exercise program. *Journal of Occupational Medicine, 17,* 240–246.

Garabrandt, D. H., Peters, J. M., Mack, T. M., & Bernstein, L. (1984). Job activity and colon cancer risk. *American Journal of Epidemiology, 119,* 1005–1014.

Gibbs, J. O., Mulvaney, D., Henen, C., Reed, R. W. (1985). Work-site health promotion: Five-year trend in employee health care costs. *Journal of Occupational Medicine, 27,* 826–830.

Haskell, W. L. (1985). Physical activity and health: Need to define this required stimulus. *American Journal of Cardiology, 55,* 4D–9D.

Haskell, W. L., & Blair, S. N. (1980). The physical activity component of health promotion in occupational settings. *Public Health Reports, 95,* 109–118.

Heath, G. W., Gavin, J. R., Hinderlith, J. M., Hagberg, J. M., Bloomfield, S. A., & Holloszy, J. O. (1983). Effects of exercise and lack of exercise on glucose tolerance and insulin sensitivity. *Journal of Applied Physiology, 55,* 512–517.

Iverson, D. C., Fielding, J. E., Crown, R. S., & Christenson, G. M. (1985). The promotion of physical activity in the United States population: The status of programs in the medical, worksite, community and school settings. *Public Health Reports, 100,* 212–224.

Jones, H. H., Priest, J. O., Hayes, W. C., Tichenor, C. C., & Nagel, D. A. (1977). Humeral hypertrophy in response to exercise. *Journal of Bone and Joint Surgery, 59A,* 204–208.

Kemmer, F. W., & Berger, M. (1983). Exercise and diabetes mellitus: Physical activity as a part of daily life and its role in the treatment of diabetic patients. *International Journal of Sports Medicine, 4,* 77–88.

Kiyonaga, A., Arakawa, K., Tanaka, H., & Shindo, M. (1985). Blood pressure and hormonal responses to aerobic exercise. *Hypertension, 7,* 125–131.

Krolner, B., Tondevold, E., Toft, B., Berthelsen, B., & Nielsen, S. P. (1982). Bone mass of the axial and the appendicular skeleton in women with Colles' fracture: Its relation to physical activity. *Clinical Physiology, 2,* 147–157.

LeBlanc, J., Nadeau, A., Richard, R., & Tremblay, A. (1981). Studies on the sparing effect of exercise on insulin requirements in human subjects. *Metabolism, 30,* 1119–1124.

Leon, A., Connett, J., Jacobs, D. R., & Rauramaa, R. (1987). Leisure-time physical activity levels and risk of coronary heart disease and death: The Multiple Risk Factor Intervention Trial. *Journal of the American Medical Association, 258,* 2388–2395.

Mack, P. B., LeChance, P. A., Vose, G. P., & Vogt, F. B. (1967). Bone demineralization of foot and hand of Gemini-Titan IV, V and VII astronauts during orbital flight. *American Journal of Roentgenology, 100,* 503–511.

Montoye, H., Smith, E., Fardon, D. F., & Howley, E. T. (1980). Bone mineral in senior tennis players. *Scandinavian Journal of Sports Science, 2,* 26–32.

Nelson, L., Esler, M. D., Jennings, G. L., & Korner, P. I. (1986). Effect of changing levels of physical activity on blood pressure ad haemodynamics in essential hypertension. *Lancet, 8505,* 473–476.

Paffenbarger, R. S., Hyde, R. T., Wing, A. L., & Hsieh, C. (1986). Physical activity, all-cause mortality, and longevity of college alumni. *New England Journal of Medicine, 314,* 605–613.

Paffenbarger, R. S., Wing, A. L., Hyde, R. T., & Hsieh, C. (1983). Physical activity and incidence of hypertension on college alumni. *American Journal of Epidemiology, 117,* 245–257.

Phelps, J. R. (1987). Physical activity and health maintenance—exactly what is known? *Western Journal of Medicine, 146,* 200–206.

Powell, K. E., Thompson, P. D., Caspersen, C. J., & Kendrick, J. S. (1987). Physical activity and the incidence of coronary heart disease. *Review of Public Health, 8,* 253–287.

Roman, O., Camuzzi, A. L., Villalon, E., & Klenner, C. (1981). Physical training program in arterial hypertension: A long-term prospective followup. *Cardiology, 67,* 230–243.

Seals, D. R., & Hagberg, J. M. (1984). The effect of exercise training on human hypertension: A review. *Medicine and Science in Sports and Exercise, 16,* 207–215.

Shephard, R. J., Corey, P., Renzland, P., & Cox, M. (1982). The influence of an employee fitness and lifestyle modification program upon medical care costs. *Canadian Journal of Public Health, 73,* 259–263.

Simon, H. B. (1984). The immunology of exercise. *Journal of the American Medical Association, 252,* 2735–2738.

Smith, E. L., Reddan, W., & Smith, P. E. (1981). Physical activity and calcium modalities for bone mineral increase in aged women. *Medicine and Science in Sports and Exercise, 13,* 60–64.

Smith, E. L., Smith, P. E., Ensign, C. J., & Shea, M. M. (1984). Bone involution decrease in exercising middle-aged women. *Calcified Tissue International, 36*(Suppl.), 129–138.

Taylor, C. B., Sallis, J. F., & Needle, R. (1985). The relation of physical activity and exercise to mental health. *Public Health Reports, 100,* 195–201.

Tipton, C. (1984). Exercise training and hypertension. In R. L. Terjung (Ed.), *Exercise and sport sciences reviews* (pp. 245–306). Toronto: The Collamore Press.

Washington Business Group on Health. (1986). *Worksite wellness media report.* Washington, D.C.: Institute on Organizational Health.

Woo, R., Garrow, J. S. Pi-Sunyer, F. X. (1982). Effect of exercise on spontaneous caloric intake in obesity. *American Journal of Clinical Nutrition, 36, 470–477.*

Yano, K., Rhoads, G., Kagan, A., & Tilleison, J. (1978). Dietary intake and the risk of coronary heart disease in Japanese men living in Hawaii. *American Journal of Clinical Nutrition, 31,* 1270–1279.

Zuti, W. B., & Golding, L. (1976). Comparing diet and exercise as weight reduction tools. *Physician and Sports Medicine, 4,* 49–57.

III CHALLENGES AND ISSUES IN WORKSITE HEALTH PROMOTION

9 Program Development and Design in Work-Site Health Promotion

Terry Mason
Johnson and Johnson Health Management, Inc.

The key to developing and designing an effective health-promotion program at the work site is planning. The successful program will be based on determining the following:

- What are the management goals and expectations for the program?
- Who will be eligible to participate in the program?
- What are the needs and interests for health-promotion programming of the participants?
- How will the program support both the goals of the management and the goals of the participants?
- What resources will be available for the program?
- How can the resources be best managed to support the program?
- What will be the best methods for promoting the program?
- What will be the best methods for educating the participants?
- How will the success of the program be determined?

The answers to these questions will supply the critical information needed for successful program planning and program promotion.

ASCERTAINING MANAGEMENT GOALS
AND EXPECTATIONS

The scientific need for disease prevention has been well established. We know that over 50% of all deaths under the age of 65 are attributed to lifestyle-related causes. Recent studies such as the Framingham Heart Study, The Alameda County Study, The Stanford Five City Project, and the MRFIT project, among others, have each strengthened the connection between negative lifestyle practices and increased risk of early morbidity and mortality.

The economic benefits of health-promotion activities, however, have been less well established. These economic benefits are frequently mentioned as being the major driving force behind management's willingness to support health promotion at the work site. The Johnson & Johnson study of the LIVE FOR LIFE® program demonstrated that companies with the program showed a significantly lower increase in inpatient hospital care for companies with their comprehensive program versus control companies without the program (Bly, Jones, & Richardson, 1986). Some information exists on the cost of poor health practices. Kristein, (1980), for example, estimates that a smoker will cost an employer between $624 and $4,611 per year in medical costs, maintenance, absenteeism, property damage, and other insurance costs. Some information exists on the cost of health-promotion activities. Brownell, Cohen, Stunkard, Felix, and Cooley (1983) estimated a cost per pound lost for a work-site group weight-control program run by lay leaders to be $4.18.

Yet, the cost of ill health or the demonstration of the cost of health-promotion activities may not be the major motivator for management to begin or maintain a health-promotion program at the work site. The program planner needs to understand the specific reasons management has for implementing a program at their work site. The criteria the management will have for judging the success of the program must also be determined.

Interviews with management, focus groups with management, and targeted surveys are methods for determining the motives and expectations of management. Reasons for implementing health-promotion efforts can include:

- Lower health care costs
- Lower absenteeism
- Increased morale
- Increased productivity
- Informed, health-conscious work force
- Recruitment of workers
- Positive pubic relations

There will probably be a mixture of driving forces for the program. It is important to establish which (and in what priority) forces drive the implementation and maintenence of any particular program. This identification allows the program planner to develop a statement of purpose to guide the planning and the evaluation of programs.

A key to success is management support. There will inevitably be competition for resources within any company, so an important first step is to affirm management's mission through a program plan.

EMPLOYEE ASSESSMENT: THE NEXT STEP

Assessing the needs and interests of the employee population that will be participating in the program is the next critical step in planning health promotion at the work site. Once management has decided who will be able to participate, assessment of this group is key to achieving participation in the program.

Methods to ascertain the health needs of an employee population include:

- Health-risk assessment
- National data bases for prevalence of disease based on age, gender, race, and geographic location
- Health-service utilization profiles
- Site demographics

The more specific the information, the better the planning tools.

Health assessments can provide the employee and the employer with a needs-based assessment. These instruments provide employees with an inventory of lifestyle practices. Biometric measures can be included to give the employee specifics on cholesterol, weight, height, blood pressure, and other health indicators. Some assessment tools can also provide an overall look at the health practices of an employee group. The cost can vary for assessment tools from the price of copying a one-page handout to over $100.00 per employee. The health assessment can also be repeated at established intervals to evaluate program outcomes.[1]

Surveys and focus-group meetings can establish employee interest. Interest survey questions can establish convenient times for programs, convenient places for holding events and classes, best methods for communicating about upcoming events/programs, acceptable costs for programs, interest in program topics, current use of community programming options, satisfaction with previous health-promotion efforts, acceptable formats for program instruction,

---

1. One source providing further information on health assessments is: *Health Risk Appraisals: An Inventory*, National Health Information Clearinghouse, P. O. Box 1133, Washington, DC 20013.

and motivation for participating in program efforts. Program planning should balance the needs and the interests of employees. Once employee needs and interests have been established, a task force can be created to determine the best strategy for implementation.

WORK-SITE AUDIT

Finally, a work-site audit can be conducted to determine what resources are available and what the needs are for environmental changes. A planner may want to find out the status or availability of the following:

Internal Resources:

Meeting rooms
Exercise space
Audiovisual equipment
Print shop
Graphic design capability
Speakers or leaders for program within the company
Possible co-sponsors within the company (i.e., employee assistance, benefits)
Help from Maintenence to set up program space
Current health-promotion offerings on site (e.g., recreation, blood pressure programs being offered by the Medical Department)

Environmental Resources:

Space to exercise
Scales available
Smoking policy
Cafeteria offerings
Vending machine offerings
Cigarettes sold on site
Work-site safety policies and procedures in effect
Room set aside for quiet time or socialization
Community offerings.

The wisest use of time and resources for program planning can be ensured by taking the time to discover availability and procedures for using work-site and environmental resources.

DETERMINING THE STRATEGY

Once an assessment of management, the program target population, and the available resources has been made, program planning can begin. Program mission, educational delivery methods, program timing, marketing approach, personnel needs, budget, evaluation, and balancing the resources will all need to be considered in determining the strategy of the health-promotion effort at the work site.

Key questions will need to be answered for each of these areas. Although the questions differ for each work site, some general ones include:

Program Mission:

What are management's goals for the program?

What are participants' goals for the program?

What are the program leadership's goals for the program?

How do these goals support or undermine each other?

How can divisive goals be resolved?

What will be the program mission?

How will this mission be communicated to management and participants?

Education Delivery Methods and Program Timing:

When will participants attend programs?

How much time will each program session take?

How would participants like to receive information?

What is the current way that participants learn new job-related skills?

What should be the reading level of materials?

What delivery mechanisms would be most appropriate for the population?

What languages and cultures are present in the participant population? How will this affect delivery?

If multiple formats are used, what should be the balance of the delivery mechanisms chosen?

Educational formats for health promotion include group programs, clinical-based programs, self-paced programs, contests, "lunch and learns," read-on-your-own, telephone delivery, and one-on-one programs.

Marketing Approach:

How will employees have access to information?

Will the program have a logo?

If a variety of community resources are to be used, how will the program maintain its own identity?

How will employees learn about one-time programs, courses, and ongoing program efforts?

How will new employees gain access to the program?

Will there be incentives? If so, how will incentives be handled in the program?

What promotional tools will be used?

How will target (e.g., high risk groups) populations be reached?

What strategies for promotion have been used successfully by the site?

How do employees currently hear about policy changes, United Way campaigns, and upcoming events?

What percentage of the budget and effort will go into marketing?

Personnel Needs:

Who will manage the program?

What skills will that person (those persons) need?

Who will lead the educational efforts of the program?

What credentials will be acceptable for leading program offerings?

What kinds of leaders can be found in the community?

What training will be needed?

What part will a work-site advisory committee play?

What part will management play in an ongoing programmatic effort?

Will staff be salaried or part time?

How will clerical work be handled?

What resources are needed for staff to be able to work effectively and efficiently?

Budget:

What resources are available for the program?

Will employees be contributing to the cost of the program? How much?

What balance will still need to be covered from another source? How will employee contributions be handled?

What is the market rate for program leaders in the community?

What will be the cost for promotion, materials, staff, outside vendors, and overhead?

How will the budget be managed?

Evaluation:

What does management see as criteria for success?

What do participants see as criteria for success?

What does the program leadership see as criteria for success?

How will success be measured?

How will results be communicated?

How will individual confidentiality be handled and ensured?

How will results be evaluated?

The answers to these questions, and questions specific to individual work-site needs, will lead to the program plan. The strategy for balancing the resources as well as meeting the mission should best fit the needs and interests of the participant and the management of a worksite.

Health-promotion development and design issues for the work site differ little from other successful program issues. The key is to determine requirements and then to design a program to meet those requirements.

REFERENCES

Bly, J. L., Jones, R. C., & Richardson, J. E. (1986). Impact of worksite health promotion on health care costs and utilization. *Journal of the American Medical Association, 256,* 3235–3240.

Brownell, K. D., Cohen, R. Y., Stunkard, A. J., Felix, M. J., & Cooley, N. B. (1983). Weight loss competitions at the worksite: Impact on weight, morale, and cost-effectiveness. *American Journal of Public Health, 73,* 1395–1396.

Kristein, M. (1980, January 9). *How much can business expect to earn from smoking cessation?* Paper presented at the National Interagency Council on Smoking and Health Workshop entitled, "Smoking and the Workplace," Chicago.

10 Work-site Health Promotion: Some Policy Issues and Concerns

Margaret A. Hamburg
Office of Disease Prevention and Health Promotion
Public Health Service
*U.S. Department of Health and Human Services**

Skyrocketing health-care costs and new understanding about disease prevention and health promotion are changing the patterns of work and health in the United States today. Over past decades, we have witnessed a shift away from problems of acute infections to chronic disease and accidents. The leading sources of morbidity and mortality in the U.S.—heart disease, cancer, stroke, accidents, diabetes, and liver disease—all have been linked strongly with risk factors affected by personal behavior. Some estimates suggest that more than 60% of premature mortality could be avoided through behavior changes, such as smoking cessation, proper diet, stress management, accident prevention, and adequate exercise (Amler & Eddins, 1987). Correspondingly, there is an increasing focus on the opportunities to improve the health of the working population and their families through work-site health promotion.

Three broad areas of activity are relevant to discussion: (1) work-site health policy; (2) national policy concerning health promotion/disease prevention, which has direct or indirect implications for the work-site; and (3) legislation. With respect to these areas, an emphasis is placed not only on identifying and exploring present concerns, but also on the need to examine emerging issues and trends related to work and health. Policy must anticipate and accommodate future concerns if effective strategies for improving health through work are to be developed and implemented. Additionally, there is a need to be alert

---

*Margaret A. Hamburg, is now a Deputy Commissioner in the New York City Department of Health.

to unintended effects of policies which may develop through health-promotion programs, with potentially adverse consequences.

PRESENT CONCERNS

A wide range of important policy-related concerns exist with regard to health promotion at the work-site. Several of the issues felt to be most critical are discussed below.

Coordination of Health-Promotion Programs

Critics of work-site health promotion have often cited competition with occupational health and safety programs as a major negative factor. While this may be an overstatement, careful attention must be given to the coordination of all health-related programs, services, and protections available at the work-site. It is critically important that health-promotion programs do not take existing resources away from fundamental occupational health and safety activities or health-care delivery services. Furthermore, health-promotion programs must not be considered as substitutes for many of the more traditional, yet essential components of occupational health. Overall, effective health promotion at the work-place must balance the responsibility of the employer to protect employees from illness and/or disability caused by work or the work environment, with the individual's personal responsibility for health. For example, offering smoking-cessation programs in the work place is not a reasonable or responsible substitute for the provision of properly ventilated working environments or a work-site free of chemical toxins that impair lung function.

On the other hand, many examples exist where proper coordination can enhance prospects for health. The work place offers an unusual opportunity for programs and policies to act synergistically. Here, again, smoking serves as a good illustration. Smoking-restriction policies in combination with easily available smoking-cessation programs can work together to address the problem of smoking in a constructive, health-promoting fashion.

In addition, the design and implementation of work-site health-promotion programs must reflect consideration and understanding of medical and other services available to employees. For instance, screening programs of various kinds may be included in work-site health-promotion programs. Such efforts are undertaken to detect health problems, many of which will require medical attention. A well-designed and successful program must make sure that appropriate follow-up opportunities are available for those in whom problems have been identified—either through the health-promotion program itself, or through access to company medical services, community services, and adequate medical benefits.

Legal/Ethical Issues

A central concern in the development and implementation of work-site health-promotion programs revolves around issues of discrimination and confidentiality. Although not often addressed in a formal manner, these matters should not be overlooked. Given the increased attention—nationally and in work sites—to the value of healthy behaviors such as not smoking, proper nutrition, weight maintenance, and exercise, those who do not follow desired patterns may be subject to social pressure and even prejudice. Although health-promotion programs should be designed to encourage positive health habits, programs must be wary of inadvertently causing discrimination against those with poor health habits. Caution should also be taken to avoid the potential for the development of coercions to participate in work-site health-promotion activities.

In addition, it is of extreme importance that the rights of workers participating in work-site health-promotion programs are protected. A strong emphasis is needed on the appropriate use of health information obtained in such programs and confidentiality of records. A successful effort will be impossible if employees fear possible job discrimination or other negative consequences.

Research Needs

At the present time, there are many gaps in our knowledge about effective strategies for disease prevention and health promotion. There is strong scientific evidence to support many aspects of prevention, and a growing body of data documenting the relationship between risk factors and health outcome. For example, it is now well established that the burden of illness produced by cardiovascular disease can be substantially reduced by modification of behavioral risk factors related to smoking, dietary cholesterol, and compliance to medical regimens to control high blood pressure (Kannel, 1978). However, in many areas there is only suggestive evidence to support health-promotion and/or disease prevention interventions. How aggressively should behavior modifications be encouraged when actual data regarding effectiveness are scant? Work-site health-promotion programs need to build on the enthusiasm for the growing body of evidence that prevention can work, while at the same time not overreaching the available data.

The scientific community must take a strong lead in exploring the nature and scope of the health benefits of disease-prevention and health-promotion approaches to provide additional information. New and ongoing research—ranging from basic science to applied—needs to be supported. Unfortunately, many of the types of research needed are expensive and time-consuming, but it is only through projects such as prospective epidemiologic studies and controlled clinical trials that the necessary answers will be obtained.

The scientific community and practitioners of health promotion in work sites and other settings must work together to identify where gaps in knowledge exist and to develop the kinds of research, evaluation, and demonstration projects needed to address those gaps.

TRENDS/EMERGING CONCERNS

Many significant changes can be identified that will affect the future of work and health. On the health side, influences include: dramatic advances in medicine and public health that have led to major improvements in health status and longevity of the American people; developments in biomedical technology, with new opportunities for the diagnosis and treatment of disease; rising health-care costs that have accompanied progress in medical and health-care capabilities; and recent insights into the importance of health-enhancing behavior and lifestyle factors that have given new perspectives on the traditional role of the health-care system.

On the work side, several important influences include: changes in the nature of the work place, technological advancements and increasing automation; the changing face of the work force with increased numbers of women, minorities, and older individuals; and an emerging emphasis on the importance of a work environment supportive of employees and their good health— both mental and physical.

It is important to examine some of these trends and to look closely at their potential impacts on policy needs and considerations.

Advances in Technology

Rapid, often stunning, changes are under way in medicine and in the work place. Future labor needs and concerns will reflect the growing role of automation. Robotics, computer systems, and other technological innovations will prove beneficial in freeing workers from hazardous job activities and/or monotonous routines, but will also raise problems of unemployment, new occupational hazards, and a different set of job dissatisfaction issues as work becomes increasingly isolated, technological and dehumanized.

Many breakthroughs from biomedical research have important implications for disease and health, both in the work place and beyond. New diagnostic and therapeutic modalities will help to reduce the burden of existing illness, and many of these techniques may be appropriate for application in work-site health programs. Also, new assessment techniques are continually increasing our ability to identify and quantify environmental, physical, and behavioral risks, thereby preventing certain health problems. For example, improved testing technologies will enable the detection of small quantities of toxic materials, and progress in molecular biology will enable scientists to make more accurate assessments of the health threats posed by exposure to such materials. Advances in genetic screening may make it possible to identify which workers have particular susceptibility to certain exposures and illnesses. Other examples of the ways in which emerging technologies may improve health in the work place include laser sterilization of equipment, improved pollution-control systems, and ergonometric furniture and equipment design (Peck, Goldbeck & Myers, 1987).

Negative potential applications of the new biomedical technologies also abound in the work place. There is a real concern that better diagnostic and screening techniques will lead to job discrimination against those with disease predispositions or asymptomatic illness.

Changing Demographics

By the year 2010, the U.S. population is expected to grow from an estimated 1985 figure of 239 million people to 283 million (U.S. Bureau of the Census, 1984). People aged 65 and over represent the fastest-growing segment of the U.S. population, and this is particularly true for elderly females. At the turn of the century, there were approximately 3.1 million elderly Americans, representing roughly 4% of the total population. By 1985, 12% of the U.S. population, or almost 29 million people, were over the age of 65. Projections suggest that as the "baby-boom" generation becomes senior citizens, there will be an estimated 65 million older persons comprising more than 21% of the total population (Fowles, 1985).

Improvements in health status will enable many of these older individuals to work past traditional retirement age; economic pressures may make this a virtual necessity for many. Retirement requirements, pension programs, and patterns of work will need to be adapted to reflect this demographic shift.

A growing number of workers are female. Between 1980 and 1985, roughly 66% of new jobs were assumed by women. Since 1980, the rate of employment among women has grown some 2.4% each year, while the rate of employment by men has grown at 0.9% (Russell, 1986). In 1985, approximately 51 million women were in the work force (U.S. Department of Labor, 1985a). Estimates indicate that in 1985, the percentage of women in the labor force was 44%, and the ratio of women to men in the work force is expected to rise markedly over the next 10 years (U.S. Department of Labor, 1985b). If present trends continue, the proportion of working women will equal that of working men after the turn of the century.

In addition to the growing numbers of women in the work force, it is important to note that a large proportion of these women will be in their reproductive years. Nearly 34 million women—about 30% of the nation's total labor force—are between 20 and 44 years of age (U.S. Department of Labor, 1986a). An estimated 80% of working women will become pregnant during their working lives (Catalyst, 1984). In 1985, 50% of all women with infant children one year of age or younger were working (U.S. Department of Labor, 1986b).

It is clear that as women take on an increasingly significant role in the work place, issues related to parental leave, child care, and reproductive health will require growing attention. Creative approaches in terms of part-time and flexible-time work schedules, job sharing, and work-at-home arrangements will need to be developed and implemented. Individual work-site policies and broader legislative plans must address the needs of women in the work place in a clear and explicit manner.

Another emerging demographic trend affecting the future of work and health involves the increased participation of minority-group members in the work force. Overall, the proportion of ethnic minorities in the U.S. population will increase because of higher birth rates in these groups, along with patterns of immigration (U.S. Bureau of the Census, 1984). Correspondingly, the health needs of these minority populations will assume new prominence, with important implications for employers.

Unfortunately, major disparities in health status exist between certain groups in the U.S. Blacks and other minorities—with the exception of Asian and Pacific island groups—suffer more ill health, disability, and premature death, as compared to the white population. The recent federal report by the Secretary's Task Force on Black and Minority Health (USDHHS, 1985) presented a disturbing picture of racial health differences across the life span. The report estimated that if mortality rates for blacks and other minorities were equivalent to those for whites, more than 60,000 deaths could be avoided each year. The task force attributed more than 80% of these so-called "excess deaths" to six specific causes, in the following order of magnitude: heart disease and stroke, homicide and accidents, cancer, infant mortality, cirrhosis, and diabetes.

It is relevant to note that for all of these conditions, there are significant associated controllable risk factors. Many of these risk factors could be addressed through appropriate work-site health-promotion activities. Clearly, lower illness and death rates from these conditions would benefit both employer and employee.

Changes in the Delivery and Financing of Health Care

In recent years, there has been a significant shift in the patterns of health care to extend beyond the boundaries of traditional settings. As the concepts of health promotion and disease prevention become accepted, the traditional delivery sites are no longer perceived as the exclusive domain for diagnosis and treatment. Increasingly, certain components of health care are being managed outside of hospitals and physicians' offices—in local clinics, community-based programs, homes, and work-sites. There is every reason to believe that this will continue to be an important trend.

NATIONAL POLICY/LEGISLATIVE CONCERNS

From a national perspective, the work site represents an opportune setting for the development and implementation of health-promotion strategies. Roughly one-half of the U.S. population is in the work force (U.S. Department of Labor, 1986a). What is more, the work-site programs have the potential to extend access to many more Americans through employee spouses and dependents, along with retiree programs.

In a general sense, national policy favors work-site health promotion. There are no legal barriers to implementing such programs, and, at least at present, the cost of many programs are deductible to employers as a business expense. However, not all employers are large enough to be able to provide employees with on-premise health-promotion programs and athletic facilities. This means that programs supported by an employer but offered elsewhere may actually be considered taxable as income to the employee. Means of ameliorating this situation are presently being investigated.

Beyond financial incentives, work-site health-promotion programs can be encouraged in other ways. The federal government stimulates and supports work-site health-promotion programs in both the public and private sectors through a variety of mechanisms. One approach is to enhance the visibility and desirability of such activities through leadership endorsement. For example, President Reagan, in a nationally transmitted teleconference, made a strong statement to executives encouraging and supporting health-promotion activities. Similarly, the Director of the Office of Personnel Management has issued a statement encouraging more activity within federal work sites.

The federal government can have a powerful influence as a role model for the development and implementation of work-site health-promotion activities. In fact, one of the very first health-promotion programs to be established and evaluated in a systematic fashion was at the National Air and Space Agency. Presently, a wide range of health-promotion activities are available to many of the federal government's approximately two-million-person work force, with activities varying considerably according to demographic, job, and work-site characteristics. Agency-wide health-promotion activities have been officially encouraged, for example, by the Departments of Defense, Transportation, and Agriculture (McGinnis, 1986).

An emerging emphasis of health-promotion activities in the federal government work place concerns so-called "smoke-free environment" policies designed to protect nonsmokers and to ban smoking except in specific designated areas (General Services Administration, 1986). Many such policies are accompanied by programs to help employees stop smoking.

Several federally sponsored national health campaigns have targeted work sites as important locales for both the dissemination of health information and the implementation of prevention strategies. Prominent illustrations include the National Cancer Institute's public-awareness campaign on cancer prevention; the Alcohol, Drug Abuse, and Mental Health Administration's campaign on depression; and the National Heart, Lung and Blood Institutes' programs on high-blood-pressure control and cholesterol education (McGinnis, 1986).

The federal government also offers technical assistance to both public and private employers through the development of educational materials and guidebooks concerning important topics in health promotion and disease prevention. For example, the Office of Health Promotion and Disease Prevention, in collaboration with private organizations and other governmental offices, has issued documents dealing with issues such as work-site nutrition (Glanz, 1986) and reducing smoking in the work place (Behrens, 1985).

In a broader sense, the government encourages work-site health-promotion activities by funding and conducting research to study health-promotion/disease-prevention strategies. There is a large body of research directed by the Department of Health and Human Services in this area. Much of this work can be applied to the work setting. In addition, considerable research activities are now being focused to specifically provide information about the effectiveness of work-site interventions. A good example of such a project is the National Heart, Lung, and Blood Institute (NHLBI)-sponsored study of effective mechanisms for high-blood-pressure control in the work place. Between 1978 and 1981, NHLBI sponsored three, three-year demonstration studies of high-blood-pressure control programs at selected sites, using a variety of screening and treatment modalities. The projects demonstrated that work-site high-blood-pressure control programs are effective in improving high-blood-pressure detection and control, as well as in reducing absenteeism (USDHHS, 1984).

A related function of the federal government in support of work-site health-promotion programs arises through the survey, tracking, and evaluation of existing program activities. The Office of Disease Prevention and Health Promotion (ODPHP) recently completed the first National Survey of Work-site Health Promotion Activities (USDHHS, 1987), a study conducted to determine the nature and extent of health-promotion activities in work sites across the nation. This kind of survey can prove very helpful in identifying both information gaps and program needs. This, in turn, can serve as a constructive guide for policy-makers. For example, the ODPHP survey clearly demonstrated an increasing acceptance and institution of health-promotion programs by major employers. However, smaller businesses were far less likely to offer health-promotion activities. No doubt this reflects both management perceptions about direct and indirect benefits of health-promotion activities and the financial incentive structures involved in establishing such programs. Perhaps by highlighting some of these kinds of issues, future policy decisions can more effectively address those areas where major needs or discrepancies exist.

CONCLUSIONS

As we move into the 21st century, it is clear that the work site will represent an increasingly important setting in meeting health needs and improving health prospects for the American people. Work-site health-promotion programs offer the opportunity to implement many important health-promotion strategies and interventions. They can offer tremendous benefits for both employers and employees—including better health, reduced health-care costs, improved productivity, and higher morale. Yet to be effective, such work-site health-promotion programs must meet several important criteria:

- They must be well designed to reflect the needs of the participants.
- They must coordinate with existing health services and/or benefits.
- They must not substitute for the provision of a safe work environment and employee protection.
- They must not be coercive or discriminatory, or compromise employee confidentiality.
- They must reflect current scientific knowledge about disease prevention and health promotion.
- They must be sensitive to changes occurring in the nature of work and health.

Work-site health policy, whether at the local or the national level, must clearly acknowledge and address these concerns. The positive potential for work-site health-promotion programs is evident, but careful and considered decision making and policy development must receive continuing high priority. This emphasis, combined with the strong leadership and enthusiastic participation of so many people working in the area of work-site health-promotion, bodes well for the future of work and health in this country.

ACKNOWLEDGMENTS

This chapter was developed from a group session entitled, ''Legislative and Policy Concerns'' held at the Academy of Behavioral Medicine Research meeting, June 14–17, 1987. Participants in this session included: Lyle Bivens (National Institute of Mental Health), Margaret Hamburg (Public Health Service), Peter Kaufmann (National Institutes of Health), Bertram Kaplan (University of North Carolina), Lynn Kozlowski (Addiction Research Foundation, Toronto, Canada), Allan Luks (Institute for the Advancement of Health), Ruth Shapiro (Pittsburgh, PA), and Robert Strauss (University of Kentucky). The contributions made by these individuals are very much appreciated.

REFERENCES

Amler, R. W., & Eddins, D. L. (1987). Cross-sectional analysis: Precursors of premature death in the United States. In R. W. Amler & H. B. Dull (Eds.), *Closing the gap: The burden of unnecessary illness* (pp. 181–187). Oxford: Oxford University Press.

Behrens, R. A., (1985). *A decision maker's guide to reducing smoking at the worksite*. Washington DC: The Washington Business Group on Health.

Catalyst (1984, June). Nationwide survey of maternity/paternity leaves. *Perspective.* New York: Catalyst staff.

Catalyst (1986, June). A corporate guide to parental leave. New York: Catalyst staff.

Fowles, D. G. (Ed.). (1985). *A profile of older Americans.* Washington D.C.: American Association of Retired Persons.

General Services Administration. (1986). Smoking regulations. *Federal Register,* 44258–44259.

Glanz, K. (Ed.) (1986). *Worksite nutrition: A decision maker's guide.* Chicago: The American Dietetic Association.

Kannel, W. B. (1978). Hypertension, blood lipids and cigarette smoking as co-risk factors for conronary heart disease. *Annals NY Academy Sciences, 304,* 128–139.

McGinnis, J. M. (1986, April 15). Statement presented before the Subcommittee on Compensation and Employee Benefits, Committee on Post Office and Civil Service, U.S. House of Representatives.

Peck, J. C., Goldbeck, W. B., & Myers, M. L. (1987). *The future of work and health strategies* (findings of "The Future of Work and Health: The National Assembly," Leesburg, VA, November 1985). Alexandria, VA: Institute for Alternative Futures.

Russell, C. (1986, October). Editor's note. *American Demographics,* 7.

U.S. Bureau of the Census. (1984). Projections of the population of the United States, by age, sex and race: 1983 to 2080. *Current Population Report,* Series P-25, No. 952. Washington D.C.: U.S. Government Printing Office.

U.S. Department of Health and Human Services; Public Health Service; National Institutes of Health; National Heart, Lung and Blood Institute. (1984). *National Heart, Lung and Blood Institute demonstration projects in the workplace high blood pressure control summary report.* Bethesda, MD.: National Institutes of Health, NIH Publication No. 84–2119.

U.S. Department of Health and Human Services. (1985). *Report of the Secretary's Task Force on Black and Minority Health* (Vol. 1). Washington, DC: U.S. Government Printing Office.

U.S. Department of Health and Human Services, Public Health Service, Office of Disease Prevention and Health Promotion. (1987). *National Survey of Worksite Health Promotion Activities: A summary.* Washington DC: ODPHP Monograph Series.

U.S. Department of Labor, Bureau of Labor Statistics. (1985a, fourth quarter). *Employment in perspective, women in the workforce,* Report No. 726. Washington DC: U.S. Government Printing Office.

U.S. Department of Labor, Bureau of Labor Statistics. (1985b, November). *Monthly labor review.* Washington DC: U.S. Government Printing Office.

U.S. Department of Labor, Bureau of Labor Statistics. (1986a, January). *Employment and Earnings*. Washington DC: U.S. Government Printing Office.

U.S. Department of Labor, Bureau of Labor Statistics. (1986b, February). *Monthly labor review*. Washington DC: U.S. Government Printing Office.

11 Interfaces Among Community-, School,- and Work-site-based Health-Promotion Programs

Richard A. Carleton
Department of Medicine
Memorial Hospital

The scientific basis of health-promotion efforts is built on the innumerable data bases which link personal behaviors to risks of disease and disease itself. Best developed are the data bases concerning cardiovascular disease risk factors and actual cardiovascular disease. Also strong are those data pertaining to health decline resulting from excess alcohol and abuse of other substances. The interactions of personal health behavior and other diseases are also clear with reference to lung disease and smoking, while other forms of cancer and other chronic disease bear less clear relationships to personal health-related behaviors. Much of this scientific base derives from epidemiologic study. Many aspects have also received support from molecular, animal model, and human-based basic research.

The scientific study of interventions desired to change human behaviors is still in its infancy, particularly in terms of demonstrating long-term outcomes. An important second scientific step has been taken through a number of cardiovascular risk-factor intervention programs targeting high-risk individuals. The Coronary Primary Prevention Trial, the Hypertension Detection and Follow-up Program (HDFP) and the Multiple Risk Factor Intervention Trial are notable examples (Hypertension Detection and Follow-up Program Cooperative Group, 1979; Lipid Research Clinics Program, 1984; Multiple Risk Factor Intervention Trial Research Group, 1982). While encouraging, the results of many of these efforts either remain unclear or have not been fully persuasive to all scientific reviewers. A third generation of trials includes some which have been completed and others which are planned. These will help to extend the scientific basis for health promotion to broader population segments. Examples include the trials to intervene with moderate, rather than severe, elevation of blood cholesterol currently being planned. The treatment of mild to moderately elevated blood pressure is supported by the subsets of the HDFP results.

As the scientific basis for intervention in any population segment is considered, the comparability of previously studied populations must also be considered. Although limited extrapolation from high-risk to lower-risk individuals may be acceptable, new scientific trials are needed to justify extension of programs to other segments of the population.

A fourth generation of research programs is now under way, seeking a firm scientific basis for health-promotion programming. Several cardiovascular risk factors and other noncardiovascular risk elements represent a continuum of risk, depending on the intensity of the factor. Thus, the risk of coronary heart disease is related in a curvilinear, accelerating fashion to increasing cholesterol levels beginning at around 160 mg/DL in the adult population For these continuously graded risk factors, mass intervention is rational, provided the risks are low. A number of population-based experiments are currently under way (Lasater, Abrams, & Artz, 1984; Blackburn, Leupker, & Kline, 1984; Farquhar, Fortmann, & Maccoby, 1985). The results of most of these are not yet available. Lessons learned to date from school-based research or community-based research targeting health outcomes can only be considered provisional. They do not yet provide the firm guidance that may exist after careful experimentation has proved that population-wide health-promotion programming is effective.

With the caveat that community- and school-based approaches, to date, have not completed their hypothesis testing, certain messages are of value to those planning and implementing work-site programs. These have been considered by the working group comprised of Richard Evans, Angela Celantonio, James Norton, Max Morton, T. George Harris, Margaret Chesney, and Richard Carleton.

Research experience in communities and schools provide guidance concerning an important general principle. Program designers and deliverers, although often scientifically based and/or academically established, and although often delivering investigatively validated program elements, will meet with greater success by involving program participants. This general principle has several implications. Careful needs assessment and appraisal of the expectations of the program participants will facilitate program effectiveness. The principles of social marketing are well applied through focus groups, random-sample surveys, and other mechanisms to target program elements at specific participant characteristics. Through collaborative planning of programs with participants, a sense of belonging and ownership will emerge. Goals will be established for the program which are simultaneously credible in the eyes of the consumer and consonant with the goals established by the program deliverers. For example, if the program is to have research elements, the participants must be aware of this goal. They should have a clear voice in establishing the goal and in attaining the goal. The objectives of the program should be established through mutual interaction to ensure mutual acceptability. Management, for example, may have an objective of reducing the burden of low back strain/ pain, while employees' concerns may center around the ravages of substance abuse. Similarly, expected outcomes of the program should not only be jointly established, but mutually agreed upon. Misleading promises of program outcomes should not be made. If reduced absenteeism is an agreed-upon goal, the program should not promise outcomes such as improved longevity. Joint program planning will also ensure involvement of interested parties (e.g., employee recreational committees) and establish mutually agreed-upon principles concerning confidentiality of data. Breaches of credibility or confidentiality, particularly when they occur in unilaterally designed programs, are likely to be fatal to the program.

An additional benefit of bilateral planning by deliverers and recipients concerns program element makeup. Many health-promotion programs at the work site and other locations focus on the small-group module for delivery of behavior-modification principles and practices. Community experience suggests that fewer than 20% of those who wish to change behavior also wish to participate in the public arena of a small group. Self-help programs, electronically mediated programs, take-home individual programs, and a wide variety of other approaches can be appraised in the original cooperative planning process.

Initial mutual agreement on suitable program outcomes facilitates evaluation. Evaluation is a requirement of all health-promotion programs. Evaluation may be part of a rigorous scientific design or, more simply, part of the mandatory process-tracking and formative-feedback loop required by any program. Policy makers, whether governmental or corporate, require data on which to base decisions concerning future allocation of resources. Incorporation of evaluation strategies and outcome criteria into the initial joint planning activities will help ensure maintenance and even generalization of successful programs while helping work-site managers and employees to avoid the potential deleterious effects of ill-conceived, ill-delivered, or ill-evaluated programs.

A second general principle from community programs may have implications for work site health-promotion programs. The work site is one component of an individual's social and behavioral ecology. The family, the church, the school, social organizations, and many other elements of the community in which we live impact changes of and maintenance of health-promotive behaviors. The principle of diffusion of attitudes and behaviors from work sites to other parts of the support networks of employees is important. It seems likely that opening work-site programs to family members not only will emerge as a priority from the previously described joint planning process, but will facilitate acquisition and continuation of new behaviors. For large work sites, endorsement, stimulation, and/or support for community-based activities may further enhance both corporate image and health-promotion effectiveness.

Health-promotion programs, as their scientific validity is proved, should not be restricted to employees of large work sites. Yet, logistics and resources indicate that these are the sites where the premier programs develop and flourish. Community experience suggests the power of networking smaller, lower-resource work sites. Indigenous resources can be shared. Cooperative health fairs can be provided. Local mini-media can be utilized. Collaborative program delivery can be actualized. Existing networks, such as the chambers of commerce, can be utilized as coordinators for efforts to reach work sites employing small numbers of individuals.

Community programs, whether targeting children in school or populations in their municipalities, suggest the power of indigenous resources in planning and implementing health-promotion programs that are culturally compatible, diversified to meet perceived needs, and designed to achieve mutually acceptable, appropriate, and achievable goals. The working group suggests these principles as being of potential benefit to future work-site health-promotion programs.

REFERENCES

Blackburn, H., Luepker, R. V., Kline, F. G., et al. (1984). The Minnesota Heart Health Program: A research and demonstration project in cardiovascular disease prevention. In J. D.Matarazzo, S. M. Weiss, J. A. Herd, et al. (Eds.), Behavioral health: A handbook of health enhancement and disease prevention (pp. 1171–1178). New York: Wiley.

Farquhar, J. W., Fortmann, S. P., Maccoby, N., et al. (1985). The Stanford five city project: Design and methods. *American Journal of Epidemiology, 63*, 171–182.

Hypertension Detection and Follow-up Program Cooperative Group. (1979). Five-year findings of the hypertension detection and follow-up program: I. Reduction in mortality of persons with high blood pressure including mild hypertension. *Journal of the American Medical Association, 242*, 2562.

Lasater, T., Abrams, D., Artz, L., et al. (1984). Lay volunteer delivery of a community-based cardiovascular risk factor change program: The Pawtucket experiment. In J. D. Matarazzo, N. E. Miller, S. M. Weiss (Eds.), Behavioral health: A handbook of health enhancement and disease prevention (pp. 1161–1170). Silver Spring, MD: John Wiley.

Lipid Research Clinics Program. (1984). The Lipid Research Clinics coronary primary prevention trial results: I. Reduction in incidence of coronary heart disease. *Journal of the American Medical Association, 251*, 351–364.

Multiple Risk Factor Intervention Trial Research Group. (1982). Multiple Risk Factor Intervention Trial. Risk factor changes and mortality results. *Journal of the American Medical Association, 248*, 1465–1477.

12 Recruitment, Attrition, and Maintenance Issues in Work-Site Health Promotion

Kelly D. Brownell
University of Pennsylvania School of Medicine

William S. Jose II
OverView Consulting

INTRODUCTION

The issues of recruitment, attrition, and maintenance are central to health-promotion programs at the work site. Several important indicators of program effectiveness are how many people can be recruited into the program, how many remain in the program, and how many maintain changes produced by the program. There are individual, group, and organizational factors which influence these matters.

Recruitment, attrition, and maintenance are significant for work-site programs as employers look to these programs as tools for health-care-cost containment rather than mere employee benefits. As an employee benefit, the employer may be satisfied that the program is positively received and available to those who care to participate. But if the employer views health promotion as a cost-containment strategy, there is a vested interest in ensuring widespread participation, low attrition, and effective maintenance. Only under these conditions will a program be cost-effective and yield a positive return on investment.

A number of factors influence recruitment, attrition, and maintenance (Fielding, 1984). These are evident at many levels, including the level of the individual (e.g., motivation), the group (e.g., social support), and the organization (e.g., smoking policies in the work place). They converge to form a complex but fascinating picture of health, behavior, and their interaction in a unique environment (the work site).

An example which crosses the weight-control and smoking areas highlights several of these issues. It demonstrates how one approach may have different

159

effects on various target behaviors, and may exert different effects on recruitment, attrition, and maintenance. A weight-loss competition was held in three banks (Brownell, Cohen, Stunkard, Felix, & Cooley, 1984). Employees within each bank who joined the program formed a team and were given a weight-loss goal, based on the number of pounds each individual needed to lose. The program generated a high level of participation (31% of the work force), attrition was extremely low (.05%), and weight losses were higher than in other work-site weight-loss programs.

A similar program was used in the smoking area by Klesges, Vasey and Glasgow (1986), but with somewhat different results. Four banks engaged in a smoking-cessation competition, with a fifth bank serving as a control (participants received a traditional smoking program consisting of small groups led by a professional). Smoking-cessation rates both after the program and at a six-month follow-up did not favor the competition banks over the control bank. However, the recruitment figures strongly favored the competition approach. In the control bank, 53% of the eligible smokers joined the program, compared to 88% in the competition banks. Therefore, if one considers the percentage of *eligible* smokers in each work site who stopped smoking, rather than the percentage who enrolled in programs, the competition produced superior results. A recent review showed other positive effects of competitions when used in community or work-site programs (Brownell & Felix, 1987).

Studies like these illustrate how recruitment, attrition, and maintenance are related, but are independent factors. One program may attract many but retain few, while another may attract few and retain many. As an example, let us say that a work site has 100 smokers. Using Program A, all 100 smokers join the program. Of these, 50 remain in the program (attrition of 50%), and 10 of the "survivors" quit smoking, yielding an overall quit rate of 10%. Using Program B, only 20 of the 100 eligible smokers join, but all remain in the program (0% attrition), and 10 quit smoking, yielding the same overall quit rate of 10%. For both Program A and Program B, 10 of the 100 eligible smokers quit, so the effect on the total health picture at the work site would presumably be the same. However, the effect might be different for these two approaches. Program A would seem to be highly visible and enticing to yield a 100% recruitment rate, so this excitement might lead to the recruitment of other people into programs targeting other risk factors. One could argue, however, that a program with such high recruitment is likely to have high attrition, and perhaps high rates of relapse, because not all participants would be highly motivated. Both attrition and relapse can cast a negative aura on a program, and relapse may have additional physiological and psychological effects on the individual (Brownell, Marlatt, Lichtenstein, & Wilson, 1986).

We do not know whether our hypothetical Program A would be better or worse than Program B after considering all of the possible effects. We present this example to show that recruitment, attrition, and maintenance are separate factors, and that only by examining all of the factors in a composite picture will we be able to evaluate the overall effectiveness of a program.

The purpose of this chapter is to provide a brief overview of what is known about recruitment, attrition, and maintenance, and to propose questions remaining to be answered. In doing so, we emphasize the importance of collaboration between researchers and the professionals who are implementing programs in work settings. It is only with good program evaluation and experimental work that the field will advance from its current state of non-theory-based, hit-and-miss programs. Conversely, researchers must be sensitive to the real-world demands imposed by work environments. In addition, creative ideas about new programs, the effects likely to be produced by programs, and the factors important to business leaders are likely to come from those with front-line experience. Therefore, the most fruitful efforts may result from a partnership between the business world and the research community.

WHAT IS KNOWN

In order to outline what is known about recruitment, attrition, and maintenance, we draw upon our knowledge of the research literature and upon the experience in work sites of the individuals who comprised the group from which this report originated (see acknowledgments). We will discuss what is known from research and what is suggested by program experience.

Recruitment

There is relatively little research on recruitment. A small amount of information is available on the individual, program, and organizational factors associated with the rate of program participation. On the individual side, Klesges, Brown, Murphy, Williams, and Cigrang (1987) found that the greater the number of fellow smokers in an employee's work area, the less likely that employee was to join a smoking-cessation program. There are studies on the characteristics of people who join clinical programs for weight loss, smoking, and exercise, but little is known about the transfer to work-site programs. On the organizational side, Klesges et al. (1987) found that work sites that had established no-smoking areas had higher participation rates in smoking programs. One program factor which appears to be associated with higher recruitment is the use of health-promotion competitions (Brownell & Felix, 1987).

From practical experience, several factors appear to be associated with the rate of recruitment. One such factor is a needs assessment to determine the employees' risk-factor profiles and their desire for various programs. It is important to note that employees' "needs," as typically determined by health-risk assessment, may not correspond systematically to their "perceived needs'' or ''desires'' for health behavior changes. From a programmatic point of view, it may be necessary to address the ''desires'' or ''wants'' first, in order to increase participation rates, and then later introduce programs aimed at the more important ''needs.'' It is thus imperative that both ''needs' and ''wants'' are identified and prioritized in order to facilitate program planning.

Establishing a planning committee with key employees and managers appears to generate visibility, commitment, and ideas for program implementation (Felix, Stunkard, Cohen & Cooley, 1985; Metcalfe, 1986). Having employees involved in the planning committee helps to create a sense of ownership. Where unions are established, it is important to have their support and representation. Management support and participation at all levels of the organization are important both to communicate that health promotion is an important business objective, not merely an employee benefit, and to model the involvement and positive orientation toward health promotion which is sought from employees.

High visibility for the program may be enhanced by announcements, posters, newsletters, and the like, and by participation of top management. Organizational factors like changing food selections in cafeterias and vending machines, removing cigarette machines, and opening exercise facilities would be logical factors to be related to recruitment. We hasten to add that these factors have not been evaluated systematically, but are worth considering in planning research and programming. This will be discussed below in the section, ''Challenges Before the Field.''

Attrition

The issue of attrition has come of age in programs for health behavior change. Especially in clinical programs, attrition was viewed in the past as something which complicated the interpretation of results; some studies did not even report the number of participants who dropped out. More recently, studies report the percentage of participants who drop out of programs, and in some cases attrition rate is one of the key dependent variables.

The interest in attrition is an important development for several reasons. First, it aids in the interpretation of reports of program effectiveness. If a program produces an impressive effect on behavior (e.g., a 20-pound average weight loss), but 9 of 10 people who join the program drop out, the effect loses some luster. Second, researchers and program planners have come to realize that attracting people into a program does not guarantee that those people will remain in the program.

A number of studies have examined the characteristics of individuals who drop out of behavior-change programs, although, again, most of what is known is from clinical studies. In these clinical studies, attrition is not typically the key variable being studied, but researchers may report correlations of questionnaire data collected prior to the program with the tendency to drop out. The variables suggested as predicting dropout are low self-motivation (although this is difficult to measure), previous history of change attempts (this is correlated with dropout in weight programs and with success in smoking programs), low education level, lack of social support, lack of early change in the program, and physiological factors (little study has been done).

There are also a number of program variables that are related to attrition. The specific components of the program are important. We know from the weight-control area that there is great variability in outcome produced by behavioral programs. The outcome depends on the quality of the programs, which is influenced by the addition of exercise, cognitive change, relapse prevention, and other factors (Brownell & Jeffrey, 1987). Such variables can be identified in any program, so we refer the reader to the chapters in this volume which deal specifically with smoking, weight control, and exercise for more information.

From the practical viewpoint, our group identified several factors which appear to be related to attrition. Having the program in a nonthreatening environment seems to be important. This provides a positive atmosphere in which participants do not feel forced to change. Using immediate prompts (telephone calls, letters, personal visits) to persons who miss one or two meetings can encourage potential dropouts to continue participating.

The convenience and quality of the program were also cited as factors which probably have an impact on attrition. Some convenience factors are the time, location, and length of the activities. In most organizations, these are easily manipulated variables. Perceived program quality for instructor-led program components is largely related to variables such as instructor enthusiasm, knowledge, and commitment to the participants. These factors can be measured with simple questionnaires.

Program cost also has intriguing effects. A moderate fee seems to decrease attrition and increase success at behavior change. However, it also appears to reduce recruitment. Research to specify the effect of these opposing influences would be of immense practical significance.

Maintenance

The standard approach to maintenance emphasizes some variety of either support groups or coping skills. Little can be gleaned from the research literature about the precise role support groups can play in the long-term maintenance of health behavior changes. In light of the high recidivism rates in many programs, it is clear that continued feedback of some type after the completion of a formal program is necessary to prevent relapse. It would be very helpful if researchers could identify the key success factors in such groups.

In the area of coping skills, a series of techniques referred to collectively as "relapse-prevention" strategies has become quite popular (Brownell et al., 1986; Marlatt & Gordon, 1985). These techniques derive from a series of principles developed primarily by Marlatt and colleagues. Programs based on these principles typically emphasize the identification of high-risk situations, distinction between lapse and relapse, methods to enhance self-efficacy, and specific behavioral and cognitive strategies for preventing and coping with lapses. Specific application of these principles has occurred in the smoking, obesity, and alcohol areas, in many cases with positive effects on long-term outcome.

One possibly important factor in program success is the use of individuals who have been successful in the program to lead others both in behavior-change classes and in support groups. It is not known specifically whether and why this is successful. It will be useful to learn from further studies whether the effect occurs because of role modeling or other factors.

Another potentially significant factor related to maintenance is the effect of simultaneous changes in several risk areas. Smoking, weight control, and exercise all seem to group naturally. Is it more effective over the long term to approach these sequentially or simultaneously?

A significant and often overlooked approach to improving long-term maintenance is to focus on aspects of the organization itself. There are two key areas to consider: organizational culture and organizational policies.

Organizational culture is a term so overused that it may appear at first to be vaporous. It is not. Organizational culture, defined by the norms which govern the appropriate and expected behavior of members, is measurable and has a profound effect on the work force. The effect of organizational expectations regarding the workaholic and alcoholic illustrate the magnitude of the impact. More subtle expectations govern acceptable food selection and tolerance for overweight. The expectation that "we may be overweight, but we're hard-drinking workaholics who really know how to party to blow off steam" can be changed to a more healthy orientation: "We take care of ourselves through exercise, nutrition, and sufficient rest so that we are peak performers both on the job and elsewhere."

More research is needed to investigate the formation and modification of organizational cultures. Particularly, we need to know how to identify the leverage points of introducing change. An organizational culture supportive of healthy lifestyle not only supports long-term maintenance of healthful behavior changes, but may also have positive effects on both recruitment and attrition.

Organizational policies can likewise have a pervasive impact on employee health behaviors. A strong no-smoking policy may be the most effective way to deal with that particular risk from the organization's point of view. Other risks, however, usually require more delicate handling.

Financial incentives built into the structure of the employee health-care plan could potentially have a great effect on a variety of employee health behaviors. Examples of what can be done include reduced premiums, reduced deductible, and/or reduced co-payment for employees at low risk on a number of risk factors. Another example: no deductible for auto-accident injuries if the insured is wearing a seat belt. It is possible to build such a program so that it costs no more than those currently in effect. Unfortunately, we receive little guidance from the research literature on this topic, and there is little practical experience to draw upon because few such approaches have been tried. Yet, this is potentially a cost-effective way of promoting health in the work place. Research on the cost/benefit of various plans would be of great assistance in this area.

Organizational policies with economic impact on the individual, such as those suggested above, are fundamentally reinforcement programs with reinforcers very distant from the behaviors they are designed to reinforce. We know little at present about the effectiveness of such reinforcers and the value of the reinforcer to the individual. Presumably academic research could be brought to bear on the design of such programs. Specifically, we need to compare the effects of symbolic and financial rewards and the application of them in the near term as immediate reinforcers for specific healthful behaviors, as opposed to applying them in the long term as in a yearly reward for low risk factors. Also, because a few large rewards may be more affordable to organizations than numerous small ones, we need to know about the effect on behavior of assured small rewards, compared to a chance of winning a larger reward.

OVERRIDING ISSUES

Several issues are relevant across the areas we have discussed here. We feel that special attention should be drawn to these issues, and that the scientific and practical implications of these factors should be the focus of more work.

The Intersection of Science and Practice

The first issue is the intersection of science and the practical demands of implementing and evaluating programs. There is a tendency for researchers to scoff at the "weak" science which characterizes the health-promotion literature. There is a paucity of randomized, controlled studies. In the minds of some academics, this justifies ignoring the area and leaving the programs to those on the front lines. Conversely, the front-line professionals can feel that researchers are hopelessly out of touch with the real world and have little to offer them in the way of improving programs. This is the classic debate between scientists and practitioners and is certainly not unique to the health-promotion field. However, it is particularly unfortunate in the health-promotion field because both academic researchers and work-site health-promotion experts stand to profit enormously from working together.

As we examine the history of work-site programs, we ask whether theory has generated programs or vice-versa. There is always talk of social learning theory, diffusion of information, and the like, but for the most part, programs appear to be driving the science. Programs are developed, and the developers then look to researchers to evaluate their programs to "prove what we know already—that they work." Such an approach relies too much on intuition, is not objective, and stands to advance the field in only small increments. Testing programs based on theory may be more difficult initially, but will lead to significant advances in the long run.

It is also important for researchers to be sensitive to the barriers to research in the work site. The ideal experimental designs are not always possible. The constructive approach is to develop new designs rather than to abandon the effort. Also, some useful ideas for research and some information related to theory testing can come from professionals with practical experience. For these reasons, we feel that there is a valuable opportunity for scientists and practitioners to interact. Both groups bring important attributes to the task of improving public health. We salute, therefore, channels such as conferences, journals, and books which can bring together the best of both worlds.

The Developmental Stages of Health Promotion

Because a number of companies now have more than five years' experience in health promotion, it is logical to consider developmental stages in the growth of organizational health-promotion programs. If there are such stages, how can they be characterized? Would knowledge of the dynamics of these programs be helpful for program implementation and success? We believe so. In discussion of programs, courses, and incentives, it is easy to lose sight of the bigger picture, the long-term organizational strategy for health promotion. Stated another way, what happens when everyone already has a headband and T-shirt?

Little is known about this currently, perhaps because work-site health promotion is a new phenomenon for most organizations, so there has not been the time nor base of experience for long-range planning. Identification of the developmental stages in organizational health promotion would definitely be valuable and would cut across all of the areas we have been discussing: recruitment, attrition, and maintenance. A large number of issues are relevant to understanding this phenomenon. The issues we have identified are discussed below and are outlined in Table 12.1.

TABLE 12.1
Issues Before the Field on Recruitment, Attrition, Maintenance, and Organizational Factors

Recruitment
1. Is 100% recruitment desirable? Broad-scale recruitment may have negative consequences if nonmotivated individuals join and subsequently drop out. What is the optimal level of recruitment and does it vary across target behaviors and types of work sites?
2. Is it preferable to target high-risk individuals and hope for large changes across small numbers (medical model) or target the general population and hope for smaller changes across large numbers (public-health model)?
3. Who should be the target of recruitment efforts? Must potential participants be approached directly, or could they be approached through "opinion leaders" within the work site who assume the responsibility of "selling" the program?
4. Can the principles of social marketing be better applied to recruitment efforts?
5. Should programs focus on maximizing participation rates or on maximizing risk reduction?

Attrition
1. Is it possible to do a systematic evaluation of dropouts after they leave a program? What is their long-term status compared to individuals who join and complete a program and to individuals who never join?
2. Can we assume that dropouts are program failures?
3. Can we standardize the method of reporting outcome results to best account for attrition? When the results of a program are reported, we must know how many people were eligible for the program, how many enrolled, how many completed, and how many of each group showed behavior and risk-factor change.
4. Can attrition be used as a valuable marker of flaws in a program? The search for subject variables to explain attrition could be considered victim blaming and may distract us from the more important task of improving our programs.
5. What short-term and long-term rewards and incentives can be used to decrease attrition?
6. Is pre-program screening a viable method for reducing attrition? Is it possible to establish screening criteria which would be reliable, valid, and equitable?

Maintenance
1. How can researchers and practitioners better separate the processes which govern initial behavior change and the maintenance of that change?
2. How can social support be manipulated or exploited to improve maintenance?
3. Is it preferable to target one risk factor in an individual or to focus on several risk factors simultaneously? Some risk factors group together naturally, such as exercise in people losing weight or weight control in individuals who stop smoking.
4. What role can peers play as program facilitators?
5. What policy changes can be made to enhance maintenance? Examples are: changing food choices in the company cafeteria, offering healthy foods in vending machines, removing cigarette machines, establishing no-smoking policies, encouraging use of stairs.
6. How can rewards for healthy behavior be implemented in a cost-effective way in the work site?

(Continued)

TABLE 12.1
(Continued)

7. Can interest and knowledge about health promotion be truly diffused into the social ecology of the work site, so that changes in behavior and health will endure beyond the initial excitement of a program?

Organizational Issues
1. What level of intervention in the organization is best to target first?
2. What occurs over time to white-collar and blue-collar workers regarding participation and behavior change?
3. What role does the organizational environment play in the long-term success of health promotion?
4. Are there stages of program change that are related to individual acquisition of knowledge and skills for preventing unhealthy behavior and developing healthy alternatives?
5. What role can organizational policies play in the long-term success of health promotion?
6. What contributes to the long-term viability of a health-promotion program when all of the "gimmicks" have been played out?
7. Should programs go for selective focus on particular risk areas or focus broadly?
8. What are the dynamics of organizational health promotion over the long run, and what implications do these developmental stages have for programming?

Knowledge of developmental stages will be valuable in planning for the long-term success of health promotion. This information will help us to continue when the honeymoon with health promotion is over and the executive "champion of the cause" is gone.

CHALLENGES BEFORE THE FIELD

We end this chapter with a call for further inquiry into a number of interesting and important areas. As is typically the case when a group assembles to address an issue, more questions are raised than are answered. This is particularly true in the work-site health-promotion area, which is in its infancy compared to many other areas in the health field.

In table 12.1 we present a list of questions categorized by the issues addressed in this chapter, with an additional category regarding organizational issues. These are the questions raised by our group as some of the most pressing issues to be addressed by both practical experience and by research. Our hope is that these will increase interest among professionals in the field and will yield more information for developing work-site programs and for understanding human behavior.

ACKNOWLEDGMENTS

This chapter is a synopsis of a group session that included contributions from the following individuals: J. J. Applebaum (Denver), Kelly Brownell (University of Pennsylvania School of Medicine), John Foreyt (Baylor College of Medicine), Thomas Garrity (University of Kentucky Medical School), Katrina Johnson (National Institutes of Health), William Jose II (Over View Consulting), Robert Klesges (Memphis State University), Lauve Metcalfe (Campbell Soup Company), Kurt Stange (Duke University/University of North Carolina), James Terborg (Oregon Research Institute), and Sharlene Weiss (Silver Spring, MD). We are grateful for their creative input. This work was supported in part by Research Scientist Development Award MH00319 from the National Institute of Mental Health and by the Weight Cycling Project of the MacArthur Foundation, both to Dr. Brownell.

REFERENCES

Brownell, K. D., Cohen, R. Y., Stunkard, A. J., Felix, M. R. J., & Cooley, N. B. (1984). Weight loss competitions at the worksite: Impact on weight, morale, and cost-effectiveness. *American Journal of Public Health, 74,* 1283–1285.

Brownell, K. D., & Felix, M. R. J. (1987). Competitions to facilitate health promotion: Review and conceptual analysis. *American Journal of Health Promotion, 2,* 28–36.

Brownell, K. D., & Jeffery, R. W. (1987). Improving long-term weight loss: Pushing the limits of treatment. *Behavior Therapy, 18,* 353–374.

Brownell, K. D., Marlatt, G. A., Lichtenstein, E., & Wilson, G. T. (1986). Understanding and preventing relapse. *American Psychologist, 41,* 765–782.

Felix, M. R. J., Stunkard, A. J., Cohen, R. Y., & Cooley, N. B. (1985). Health promotion at the worksite: A process for establishing programs. *Preventive Medicine, 14,* 99–108.

Fielding, J. E. (1984). *Corporate health management.* Reading, MA: Addison-Wesley.

Klesges, R. G., Brown, K., Murphy, M., Williams, E., & Cigrang, J. (1987). *Factors associated with participation, attrition, and outcome in a smoking cessation program at the worksite.* Paper submitted for publication.

Klesges, R. C., Vasey, M., & Glasgow, R. (1986). A worksite smoking modification competition: Potential for public health impact. *American Journal of Public Health, 76,* 198–200.

Marlatt, G. A., & Gordon, J. (1985). *Relapse prevention.* New York: Guilford.

Metcalfe, L. L. (1986). Campbell Soup Company's turnaround health and fitness program. *American Journal of Health Promotion, 1,* 58–68.

13

Cost-Benefit and Cost-Effectiveness Analysis in Work-Place Health Promotion Programs

Jonathan E. Fielding
Schools of Public Health and Medicine
University of California, Los Angeles
and
Johnson & Johnson Health Management, Inc.

Employers have adopted employee health-promotion programs for a number of reasons. Most common among these are the desire to improve employee health, to enhance employee morale and job satisfaction, to reduce the rate of increase of health-related expenses (Fielding, 1984; Fielding & Piserchia, 1989).

The decision-making process for launching health-promotion activities at the work place has not been systematically studied. Usually, the programs coalesce from a number of disparate activities, started at different times and often under different auspices. As programs grow, the associated dollars usually reach the level that invites the question, What are we getting for that expenditure? This chapter suggests approaches and related pitfalls in trying to answer that question.

Most employers prefer to be able to measure both investment and returns in dollars. Outside observers often assume that the former is well known, and that the latter is attainable with minimal effort. However, often the cost of programs is not well documented. Frequently, they are buried in a number of different cost centers which may not be specifically identified. In addition, the internal personnel resources and related overhead to plan, develop, and administer the program over time are not separately budgeted, nor is time cost accounted to permit accurate calculation of these costs. In organizations where activities are provided on employer time, it is rare that the number of "lost" hours is tallied and given a monetary value.

Therefore, program cost, the common denominator required for both cost-benefit analysis (CBA) and cost-effectiveness analysis (CEA), is often unavailable. However, almost all employers possess the expertise and experience to accurately cost a program by aggregating all related costs into a defined cost center or systematically identifying the costs in different cost centers and summing them.

Much more problematic for employers is trying to determine the results of health-promotion programs they have supported. The essence of both CBA and CEA is to be able to relate costs and effects, and to use this information to channel resources in ways to maximize overall effects. Dollars are always limited to a greater degree than potential opportunities for spending (investment). Many employers handle the problem of competing needs by establishing a "hurdle rate" and only investing in those programs that achieve a predetermined hurdle rate, such as a 16% annual return on invested dollars.

CBA is the tool that should provide the information necessary to determine the return on investment of employer dollars and to permit a better comparison between its returns and those of other potential investment opportunities. However, at the time a decision for initial investment is required, the only information generally available is a set of assumptions that translate into an imputed dollar benefit, or data provided by other employers or from published evaluations.

Uncertainty about benefits is the only certainty in management's efforts to make decisions among competing uses of funds. Therefore, if an investment decision is made strictly on economic considerations, it is a choice between competing projections, virtually all of which have some tenuous underlying assumptions. For example, projections for return on investment in a new piece of equipment designed to decrease manufacturing costs depend on assumptions about relative efficiencies, maintenance costs, personnel costs to run the machine, training costs, period of useful life, utilization level, etc. The difference between analysis of health-promotion programs and other investment options is not the use of projections rather than a guaranteed stream of returns but possible differences in: (a) availability of credible sources for benefit levels; (b) strength of the assumptions underlying the projections; and (c) comfort of decision makers with the nature of the investment.

Senior management of most employers has much more familiarity with decisions about plant, equipment, training programs, etc. than about health-promotion programs, and this may influence allocation of investment dollars. Some factors that may influence them to decide to make a health-promotion program investment include: (a) presence of a significant number of health-promotion activities already, with a positive employee response; (b) major competitors with such programs and large programmatic benefits claimed by these competitors; (c) high level of concern with the quality and availability of human capital; (d) positive personal and family experiences of decision makers with health maintenance and risk-reduction activities, including successful health behavior change; and (e) high credibility of the advocates for the program. For example, a medical director who has served as the personal physician of key senior executives may have the credibility to strongly influence, positively or negatively, management decisions about a health-promotion expenditure.

Although having strong evidence of a high return on a health-promotion dollar investment would increase the chances of a positive investment decision, lack of certainty about dollar benefits may not constitute a strong barrier to investment. Decisions about health-promotion programs, like many other senior-management decisions, are not necessarily made "on the numbers," but are strongly influenced by many subjective factors.

There is an existing body of evidence which implies positive ratios of benefits to costs for a variety of work-site health-promotion programs (Fielding, 1984; Warner, Wickizer, Wolfe, Schildroth, & Samuelson, 1988). However, most of the published CBA literature in this area suffers from a number of methodological problems. This literature has been reviewed by Warner et al. (1988), who found limited evidence supporting claims of dollar savings resulting from health-promotion programs at the work place.

There are two types of published results that include CBA. One is program reports by employers themselves in the trade press or in non-peer-reviewed journals. Although many of these reports indicate a high rate of return, generally from reductions of rates of increase in health-benefit costs and, less frequently, absenteeism reductions, the reports generally do not indicate a research design that would meet academic standards or provide sufficient information on methods to judge the validity of the claim. The second type, the peer-reviewed literature, also suffers from many design and method problems. Very few of the studies employ randomization, and many do not employ control groups. Single-group time series are common, but often lacking sufficient description of possible confounding variables to support observed changes being causally linked to the health-promotion program. Control groups, when used, are often significantly different from the intervention group with respect to both sociodemographic and health variables at baseline. Analysis of covariance, while an excellent tool in many such situations, cannot account for all sources of potential systematic bias. External validity is often also compromised by the lack of an adequate description of the intervention, the use of nonstandard measures for some variables, and the limited range of sociodemographic profiles of workers studied, as well as a narrow range of job types.

Difficulties exist in measuring both costs and benefits. In theory, costs of a health-promotion program perhaps should include adverse effects of the program: those employees who leave, rather than be subjected to increased pressure to stop smoking, injuries due to exercise initiated as part of an employer-sponsored program, increased stress on those who feel pressure to change adverse health habits but are unable to do so, etc. However, these categories are difficult to identify, measure, and assign a monetary value. In parallel fashion, many benefits usually go uncounted, such as improved recruitment, enhanced retention, and, most importantly, improved productivity. Decisions on what to measure are usually influenced more by the availability of data than what outcomes are most important to the organization. The biggest problem in CBA is the need for all benefits to be measured in dollars, despite the absence of acceptable methods. Even in instances where management unanimously agrees that observed changes in morale, job satisfaction, satisfaction with working conditions, etc., represent dollar savings to the organization, no consensus method currently exists for this translation.

Despite these many limitations, a series of CBAs published in peer-reviewed academic journals is accumulating for health-promotion interventions sponsored by employers. Some evaluations have included single-focus interventions, such as hypertension control and smoking control (Alderman, Madhavan, & Davis, 1983; Fielding, 1984). Several evaluations of more comprehensive multicomponent programs have also been published. For example, programs sponsored by Blue Cross and Blue Shield of Indiana (Gibbs, Mulvaney, Henes, & Reed, 1985), and Johnson & Johnson (Bly, Jones & Richardson, 1986) have reported savings in the costs of health benefits in intervention groups compared to controls. Although these results may increase the confidence of decision makers in other companies about the value of investing in work-place health promotion, there is often concern about the degree to which results obtained in one organization can be extrapolated to others.

Another approach to estimating benefits is to estimate health benefit costs attributable to lifestyle habits and project possible savings based on likely risk reductions from a health-promotion program (Knight, Goetzel, & Fielding, 1988; Leviton, 1989). Given specific employee population demographics and projecting an associated pattern of health risks, the likelihood of risk-related adverse events (e.g., heart attack, stroke, serious motor vehicle injury) over a 5- or 10-year period can be forecasted. Employer costs for these events can be estimated from either available literature or, preferably, from an analysis of the employer's own claims experience. Degree of risk reduction can be estimated from experience with a well-established program or can be estimated using the mean or median result of all published program evaluations which meet predetermined criteria for study design and methodology. By incorporating these data into a computer model, savings from instituting a health promotion program can be estimated and compared with program costs to derive a cost-benefit ratio. Benefit assumptions (program effectiveness, discount rate, employee turnover, etc.) can be varied to provide a broader perspective on the likelihood of achieving benefits that exceed costs.

Answering those questions that are amenable to CEA usually requires fewer and more easily defensible assumptions than is usually required by a CBA. CEA asks simpler questions, such as which is a better investment, intervention A, B, or C, etc. Unlike CBA, in CEA outcomes can be measured in any standardized health-relevant units, such as reduction in mm Hg for blood pressure, percentage increase in max VO^2 for fitness, or pounds lost. A straightforward application of CEA might be how a company should best spend a set sum of dollars on a weight-management program. Outcomes might be measured in total pounds lost at a given time during the program using alternative interventions. For example, Brownell compared weight-loss program costs between professionally led weight-control groups, lay-led groups, and a competition among three bank work sites. The cost per 1% reduction in weight after 12 weeks was $25.14, $8.28, and $2.93, respectively (Brownell, Cohen, Stunkard, Felix & Cooley, 1984). Based on this analysis, the highest health benefit per dollar spent was thus work-site competitions.

However, using such analyses to make health-promotion program decisions has important limitations:

1. The experience at one point in time may not reflect the relative effectiveness upon longer follow-up. Recidivism may not be equal in all types of interventions, and relative levels of cost-effectiveness among interventions may change over time.

2. A key variable, but one often not considered in these analyses, is what proportion of the total population risk has been ameliorated. The most cost-effective intervention may only lead to 10% of the at-risk population participating, versus a much higher participation rate in interventions which overall are less cost-effective. Thus, population effectiveness (i.e. impact on the entire population at risk) may be more important than efficiency.

3. The risk levels of the at-risk population exhibit considerable variation. One intervention may be more effective with respect to the grossly obese, while another may be more effective with the moderately obese population. It may be, for example, that to reach both populations requires use of both interventions. Looking at cost-effectiveness alone obscures these important programmatic considerations and may lead to decisions at odds with overall program objectives.

4. Cost-effectiveness of alternative interventions may vary in the same population, based on prior experience. An intervention which is the most cost-effective intervention the first time it is presented to a population may turn out to be less cost-effective the second time, after the most willing employees have already participated.

As an evaluation tool to help make informed resource-allocation decisions, CEA provides very limited help in deciding between different health-risk targets. For example, how can cost-effectiveness results be compared for weight management, fitness, and stress management? Methods to compare the health value of decrements in different health-risk variables are not well developed.

Only where a strong body of epidemiologic studies exists to quantify the relative risk reduction of different risk factors with respect to a common and well-defined health outcome, can some comparison of cost effectiveness be made among programs targeting different risk factors. For example, if the desired outcome was a reduction in myocardial infarctions among white males over 10 years, use of generally recognized multiple logistic equations to quantify myocardial infarction risk would permit comparing the cost-effectiveness (with respect to prevention of myocardial infarction) of results of hypertension-control, cholesterol-reduction, and smoking-cessation programs. However, the overall cost effectiveness of these alternative programs could not be assessed because each has a number of other health benefits in addition to the effect on myocardial infarction risk. Smoking cessation, for example, has a major impact on risk of many cancers, chronic lung disease, low birthweight births, and so on.

Synergy between different health-promotion interventions is often assumed and examples have been reported, but there is no agreement on how to reproducibly quantify the synergy in considering cost-effectiveness. Determining synergy or negative interference between programs requires trying each alone, as well as all combinations.

Although this chapter has principally addressed CBA and CEA from the standpoint of the employer who is the principal funding source for most health-promotion programs at the work site, other perspectives also merit consideration. An employee who plans to leave his current employer in a year may have a strong reason to take advantage of a free or inexpensive risk-reduction program at the work site, even though the majority of the potential health benefits will not accrue to his current employer. A high-risk employee who has had difficulty making and sustaining health-promoting behavior changes may have the best chance of making lasting changes if a high-risk screening and personnel-intensive counseling and education approach is implemented, while an ecological intervention might be much more cost-effective or cost-beneficial for the entire work-site population.

All or most employer expenses for health-promotion programs are deductible for both federal and state taxes. Tax deductibility implies agreement by our elected representatives that such programs are in the *public* interest, as tax deductibility is a strong positive program incentive. From the public standpoint, a health-promotion program is beneficial, regardless of the turnover rate of an individual employer, because the benefits of the program will accrue to society in terms of improved productivity and reduced illness.

However, an employer with a high turnover in the work force is unlikely to initiate a program if only a small fraction of the total benefit will provide a return on that investment. Investment in programs which will lead to greater longevity, but not significantly affect productivity or health costs during active employment or retirement (when Medicare generally becomes the primary health-insurance carrier, from age 65 on) is much less attractive from an employer perspective than from a societal perspective. By contrast, programs such as a smoking-cessation program may lead to significant savings to employers, but could increase government Medicare and Social Security payments. One estimate is that each male light smoker who quits at age 45 when mortality rates for quitters still are close to the rates of those who have never smoked, will lead to $204 to $2,745 in additional spending from the Social Security Trust Fund for Medicare payments over the life of an individual. (Wright, 1986) While this estimate is speculative, it illustrates possible asymmetry between employer and governmental perspectives on program costs versus benefits.

The allure of CBA and CEA in helping employers decide on investment in health-promotion programs surpasses their current value to the decision-making process. Nonetheless, since the primary stimulus in many business decisions is return on investment, increased use of whatever results emanate from these types of studies can be anticipated. Considerable additional work is needed to reach consensus on methodologies for measuring costs, monetary benefits, and health benefits in health promotion interventions in the work place. Employer education on both the appropriate uses and significant limitations of these studies also deserves priority attention. Even when properly interpreted, results of CBA and CEA studies should be generalized with caution, given the heterogeneity of work forces, of the nature of the work, and of the sociocultural environment at different work sites.

REFERENCES

Alderman, M. H., Madhavan, S., & Davis, T. K. (1983). Reduction of cardiovascular events by workside hypertension treatment. *Hypertension, 5* (6, Pt. 3), v138–v143.

Bly, J. L., Jones, R. C., & Richardson, J. E. (1986). Impact of worksite health promotion on health care costs and utilization. *Journal of the American Medical Association, 256,* 3235–3240.

Brownell, K. D., Cohen, R. Y., Stunkard, A. J., Felix, M. R. J., & Cooley, N. B. (1984). Weight loss competitions at the work site: Impact on weight, morale and cost-effectiveness. *American Journal of Public Health, 74,* 1283–1285.

Fielding, J. E. (1984). Health promotion and disease prevention at the worksite. *Annual Review of Public Health, 5,* 237–265.

Fielding, J. E., & Piserchia, P. V. (1989). Frequency of worksite health promotion activities. *American Journal of Public Health, 79,* 16–20.

Gibbs, J. O., Mulvaney, D., Henes, C., & Reed, R. W. (1985). Work-site health promotion. *Journal of Occupational Medicine, 27,* 826–830.

Knight, K. K., Goetzel, R. Z., & Fielding, J. E. (1988). Unpublished Lifestyle Claims Analysis.

Leviton, L. C. (1989). Can organizations benefit from worksite health promotion? *Health Services Research, 24,* 159–189.

Leviton, L. C., Andiorio, J. M., Barkman, M., Curtis, E. C., Dinman, B. D., Epstein, L. H., Hollenbeck, B. B., Kuller, L., Longest, B. B., McDonough, L. R., Michaels, L. G., Russell, P., Stackhouse, P. A. Jr., Trump, R., & Judd, R. M. (1985). *Worksite health promotion and its status in southwestern Pennsylvania.* Pittsburgh: Health Policy Institute, Graduate School of Public Health.

Warner, K. E., Wickizer, T. M., Wolfe, R. A., Schildroth, J. E., & Samuelson, M. H. (1988). Economic implications of workplace health promotion programs: Review of the literature. *Journal of Occupational Medicine, 30,* 106–112.

Wright, V. B. (1986). Will quitting smoking help Medicare solve its financial problems? *Inquiry, 23,* 76–82.

14 In Hot Pursuit of Health Promotion: Some Admonitions

Marshall H. Becker
School of Public Health
The University of Michigan

The past two decades have witnessed the rapid growth of health-promotion activities, programs, and related providers as a major force in the American health-care system (Kaplan, 1985). We have become an extraordinarily health-conscious society, expending huge amounts of time, money, and effort in attempts to preserve our health (Barksy, 1988). While the overall conceptual strategy (i.e., altering individuals' lifestyles with a view to preventing premature morbidity and mortality) seems clearly laudable, serious questions have arisen about many of the operational tactics recommended for achieving this goal. I will discuss some problematic aspects of the health-promotion movement in three major areas: soundness of our knowledge base; untoward consequences of well-intentioned changes in behavior; and health promotion's focus on the individual.

KNOWLEDGE BASE

What do we know, and when do we know it? The public is regularly counseled to avoid a multiplicity of conditions (e.g., obesity, lack of exercise, high-stress environments) and substances (e.g., cigarettes, cholesterol, coffee) which are held to be serious threats to health. Because "clean living" (from the health-promotion perspective) often requires individuals to undertake and maintain difficult and pervasive alterations in their lives and work, it is necessary to consider with what degree of scientific certitude our warnings and recommendations can be made.

Even a superficial review of the relevant literature reveals extensive and profound disagreements among health professionals concerning the influence—on personal health and longevity—of obesity, cholesterol, exercise, Type A personality, early-detection screening, consumption of alcohol or coffee, and so forth; disagreements which result in the offering of confusing and contradictory advice. As Levin (1987) notes, "The approximate half-life of a so-called health fact is four years. This leads to substantial public confusion about what and who to believe, thus progressively diminishing the effectiveness of health messages" (pp. 57–58).

In reality, I believe that our knowledge base does not yet provide sufficient support for many of the things we now suggest that people can do for themselves (Eisenberg, 1977), and, further, that "with the noblest of motives we continue to embrace new preventive measures long before we can claim success with previous ones" (Fihn, 1987, p. 2417).

Examples of the knowledge-base problem abound (e.g., in the area of routine screening, see: Berwick, 1985; Bloom & Soper, 1986; Langone, 1985; Love, 1985; O'Malley & Fletcher, 1987; Skrabanek, 1988a, 1988b;). For illustrative purposes, it is useful to focus on health-promotion recommendations related to eating, since: (a) it involves a behavior necessary to existence; (b) its patterns are intimately connected with our personal and social behaviors (making alterations quite difficult to initiate and sustain); and (c) it is alleged to be an important causal factor in a variety of diseases and conditions.

When is a person "overweight"—and when is being overweight deleterious to health? Professional debate about this issue has been prominently featured in the media (Eckholm, 1985). Can eating certain types of foods either cause, or reduce the risk of, cancer? After an extensive review of the literature, Underwood (1986) concluded that "As yet, most of the evidence is only suggestive that altering your dietary intake patterns in the direction recommended by expert groups will reduce risk of some kinds of cancer" (p. 210). At the request of the 97th Congress, the U.S. Office of Technology Assessment (1982) undertook a study of cancer risk, and determined that "there is no reliable indication of exactly what dietary changes would be a major importance in reducing cancer incidence and mortality" (p. 76).

It is instructive to examine in detail the epidemiological data underlying current exhortations with regard to the benefits of reduced cholesterol levels. For example, Grundy et al. (1982) listed 10 references as the best available evidence to support the American Heart Association's recommendation that the amount of dietary cholesterol be limited to 300 mg/d. In examining these citations, Reiser (1984) discovered that almost all of them were either irrelevant or actually antithetical to the recommendation, and concluded that "the Rationale (Grundy et al., 1982) is not a logical explanation of the dietary recommendations but an assemblage of obsolete and misquoted references."

A review (Stallones, 1983) of seven important epidemiologic investigations of possible relationships between development of ischemic heart disease and intake of cholesterol (or fat, or calories) found the only observed differences between those who become ill and those who remain well to be in the "wrong" direction. There are many other major studies that did *not* find cholesterol levels to be related to mortality due to heart disease (Ahrens, 1979; cf. Avogaro, 1985; Coronary Drug Project Research Group, 1970; Keys, 1970; Medical Research Council, 1975; Neaton et al., 1981). Much cogent criticism of the procedures and recommendations of the National Institutes of Health Consensus Conference (1985) on Lowering Cholesterol to Prevent Heart Disease has emerged from eminent cardiovascular researchers; Ahrens (1985) asserts that the consensus statement "promises benefits without giving the evidence to back up that promise. By failing to emphasize what we do not know, the statement sweeps these weaknesses under the rug, as if they were trivial" (p. 1087).

Recently, Taylor, Pass, Shepard, and Komaroff (1987) constructed a model to estimate potential benefits of a lifelong program of dietary cholesterol reduction in terms of increased life expectancy. They calculated that persons aged 20 to 60 years who were considered to be at low risk (where risk is defined relative to the person's blood pressure, smoking history, and high-density lipoprotein level) might expect to live only three days to three months longer, while those at high risk would gain only 18 days to 12 months. Kaplan (1985) also calls attention to the fact that, in each of five studies linking dietary changes to reduced mortality from heart disease, such reductions were balanced by *increases* in deaths from other causes; life expectancy thus remained unchanged (for further critical comment with regard to cholesterol reduction, see Palumbo, 1988).

Thus, careful review of the data suggests that we have oversold the health benefits that public compliance with our eating-related advice is likely to yield. In a recent article designed for a lay audience, Dr. Jules Hirsch, the Chairman of the 1985 National Institutes of Health Consensus Conference on Obesity, wrote: "The truth is that we do not have final and definitive data on the role of diet in preventing chronic diseases, ameliorating behavior disturbances or prolonging life" (Hirsch, 1986, p. 10).

Similar implications can be drawn from close examination of empirical work providing the basis for many of our other health-promoting recommendations. This should not be surprising; personal control over health is often substantially limited by biology, heredity, social class, environment, culture, and chance (Becker, 1987; Gunning-Schepers & Hagen, 1987). The mass media, well aware of society's obsession with health matters, have provided the public with blow-by-blow coverage of putative newly discovered risks to health, heightening the problem by attributing an unjustified degree of certainty to each new development or recommendation (Becker, 1986). The situation is exacerbated by a scientific community which rushes tentative findings into print (Lipton & Hershaft, 1985; Winsten, 1985).

Levin (1987) has observed that "No standards for truth in advertising have been promulgated with regard to the proposed benefits of nutrition, exercise, or stress management" (p. 58). We cannot, of course, ever be absolutely certain of the efficacy of our recommendations. However, I believe we can (and should) exercise far more caution and restraint than has been exhibited by our performance to date, especially because: (a) we are generally not dealing with health emergencies, but, rather, with factors which *may* alter the risks of incurring eventual untoward events; (b) we often alter (or reverse) our advice in any case; and (c) as representatives of the scientific community, we provide the only checks and balances in this area.

UNTOWARD CONSEQUENCES

I also believe that there exists a tendency to ignore (or be less than candid about) the potential "down side" of health-promotion recommendations. This represents a higher-order ethical question than that involved in offering merely inefficacious advice; it evokes a fundamental concern found in the Hippocratic oath: *primum non nocere* ("first do no harm"). A growing body of literature suggests that many of our risk-reduction suggestions carry substantial risks of their own.

Little is known about the possible adverse physical and psychological consequences of major dietary modifications and restrictions. Hirsch (1986) observes that "obsessive concern over food choices may well be more harmful in the long run than any item we habitually eat" (p. 10). The great preponderance of individuals who attempt weight loss either fail initially or are unable to maintain the reductions (Stunkard, 1978), leading both to constant unhealthful "up-and-down" fluctuations in weight, and to negative self-appraisal. Over-concerned parents have placed their infants on overly stringent diets, and fear of obesity is so strong that it has affected fourth-grade girls, 80% of whom are afraid of gaining weight and who feel that they weight too much (Barksy, 1988).

With regard to drug treatment to lower cholesterol levels, Taylor et al. (1987) provide several citations in support of the statement that "the effect of cholesterol reduction on mortality from causes other than ischemic heart disease is an unsettled issue" (mortality due to cancer being the chief concern). Virkkunen (1985) notes that patients in the Coronary Primary Prevention Trial who had low cholesterol levels also had higher than expected rates of accidental and violent deaths, and reports that other research has shown impulsive homicidal and suicidal behavior to be connected with low cholesterol levels. (Why should this be? Does compliance with a low-cholesterol diet make one wish to commit murder—or to *be* murdered? Persons who have attempted weight loss or other diets may find both hypotheses plausible.)

There are examples aplenty beyond the case of eating behaviors. For instance, it is estimated that 38% of runners are injured each year (Koplan, Powell, Sikes, Shirley, & Campbell, 1982). Of the 13% of cyclists who experience accidents each year, 62% are injured, one-third of whom seek medical care (Kruse & McBeath, 1980). About 45% of persons engaging in racquet sports are injured during their playing history (Koplan, Siscovick, & Goldbaum, 1985). Thus, considerable morbidity and mortality are generated while seeking better health through exercise (cf. Kraus & Conroy, 1984). Koplan, Siscovick, & Goldbaum (1985) comment that we must "provide the public with information that presents a full and balanced view of exercise, namely, its benefits and risks. . . . If we are to maintain professional credibility, we must assess exercise risks with the same rigor that we demand of benefit analysis" (p. 194).

Perhaps our most seemingly innocuous and intuitively attractive approach to prevention would be mass-screening programs aimed at early detection of various conditions. Here, too, however, there are considerable dangers. For example, we advocate routine occult-blood screening for colo-rectal carcinoma. While data do not appear to be available documenting actual reductions in morbidity or mortality for persons who are screened, there *is* evidence "that significant harm may result from the test's widespread application, due to the need for extensive endoscopic and radiological investigations in persons with a positive result (the vast majority of whom will subsequently be found not to have cancer) . . . the risks of occult-blood screening could outweigh the benefits of any age group" (Frank, 1985, p. 25).

Turning to large-scale (20 million people) diagnostic screening for coronary artery disease in an asymptomatic population, Bloom and Soper (1986) suggest that more than 8,000 individuals experience major complications, and nearly 2,000 die from diagnostic testing and therapy (with about one-half of those complications occurring to persons without disease). We have long recommended that women regularly engage in breast self-examination (BSE), but the procedure turns out to have poor sensitivity and specificity, and the psychological effects of learning and performing BSE are not known. Reviewing this literature, O'Malley and Fletcher (1987) conclude that "many questions require scientific examination before this procedure can be advocated as a screening test for breast cancer" (p. 2197).

But perhaps the most harmful side of our health-promotion appeals (possibly because it is so diffuse, relentless, and insidious) has been the creation of what Thomas (1983) has termed an "epidemic of apprehension," wherein our almost-daily frightening declarations warn that danger lurks in every aspect of our lives: "the air we breathe, the water we drink, the food we eat, the homes we live in, the substances we touch, and the work we do" (Feinstein & Esdaile, 1987, p. 113). In efforts to encourage society to adopt our suggestions, we often sprinkle liberal quantities of "fear arousal" as a kind of motivational seasoning on our messages; this frequently generates considerable concern, but little subsequent behavior change, with the net effect of converting "persons at risk" into "anxious persons at risk" (Job, 1988). A very substantial and influential health-promotion industry has evolved, which both promotes and capitalizes on these circumstances—and "the widespread commercialization of health and the increasing focus on health issues in the media have created a climate of apprehension, insecurity, and alarm about disease" (Barsky, 1988, p. 414).

In sum, it seems paradoxical that, while medical care delivery (whose effectiveness is either established or scheduled to undergo clinical trial) is fairly closely scrutinized by government agencies, third-party payers, professional standards, etc., the health-promotion industry is essentially unregulated. Levin (1987) points out that there is "no credible, visible, or sustained resource in the public-health establishment that monitors health promotion, critiques its limits and hazards, and offers the necessary programs of public protection" (p. 58).

INDIVIDUAL FOCUS

Currently, health-promotion policies and recommendations are focused primarily on behavioral intervention at the level of the individual. Three assumptions appear to underlie this approach: (a) personal health-related behaviors are discrete and independently modifiable; (b) anyone can decide to alter his/her behavior, and then go on to successfully do so; and (c) everyone has a personal responsibility to "live well" through self-discipline and behavior modification (Coreil & Levin, 1985; Wikler, 1987). While no one would deny that most of us would be healthier if we took better care of ourselves, these assumptions do not fit very well with what we know about the major determinants of health and the prevention of illness (Brown & Margo, 1978).

First, health habits are acquired within social groups (family, peers, the subculture), are often supported by powerful elements in the general society (e.g., advertising), and have proven to be extremely difficult to change (Etzioni, 1972). Second, for most people, personal behavior is not the primary determinant of health status (Freudenberg, 1981; Slater & Carlton, 1985), and it will not be very effective to intervene at the individual level without concomitant attempts to alter the broader economic, political, cultural, and structural components of society which act to encourage, produce, and support poor health (Alonzo, 1985; Milio, 1981). As Carlyon (1984) observed when commenting on jogging, "I'm not sure how well it works for the unemployed, the unskilled, inner-city welfare mothers, Asian-Pacific refugees, the poor, the handicapped, and the dispossessed among us" (p. 29).

Third, the behaviors to be modified are only *risk* factors. Engaging in risky behaviors may, for some individuals, increase the risk of undesirable (albeit low-probability) health outcomes, but avoiding them is certainly no guarantee that untoward health events will not occur; heredity, accidents, etc., play very important roles in determining health and longevity. It is sobering to contemplate the results of a project that pooled data from six major prospective investigations that attempted to find predictors of health and disease (Pooling Project Research Group, 1978). During the established period of 10 years, only about 10% of those men who had two or more risk factors developed heart disease, and about 60% of those who developed heart disease had either no risk factors or only one known risk factor.

Fourth (and, in my opinion, most disturbing) is the translation of individual responsibility for health into a new morality by which character and personal worth are judged. "Being ill" is redefined as "being guilty." The obese are stigmatized as "letting themselves go." Smokers "have no will power." Nonaerobicists are "lazy." With intense zeal, we pursue "the Holy Grail of wellness" (Carlyon, 1984), and the as-yet unproselytized are treated as sinners. Gillick (1984) writes that in the 1970s, exercise attracted a wide audience "not because of the justification for its use provided by the scientific community, but because of the appeal of upright living as a means to personal and social redemption" (p. 369). Health has acquired tremendous importance in American society; in cross-national surveys, 46% selected "good health" as their greatest single source of happiness (over alternative responses that included "great wealth" and "personal satisfaction from accomplishments"; Harris, 1987).

What happens to us when health becomes our paramount value? Advocates of health promotion and "wellness" claim to be striving for "self-actualization" and "personal fulfillment"; but theologians and philosophers have generally agreed that to attain such fulfillment, one must make a commitment to something beyond oneself—quite the opposite direction from an emphasis on personal risk factors and lifestyles. I fear that, as it is currently practiced, health promotion fosters a dehumanizing self-concern that substitutes personal health goals for more important, humane, societal goals. It is a new religion, one in which we worship ourselves, attribute good health to our devoutness, and view illness as a just punishment for those who have not yet seen the Way—a view that evokes notions of social Darwinism and the "me generation" (Gurin, 1984; Stein, 1982). For society's (and humanity's) sake, we must begin to turn our concerns outward. If, indeed, avoiding health risks will buy us a few more years of life, we should be worrying about the quality of the society and environment in which those years will be spent.

CONCLUSION

It would appear that, for all its virtues, the health-promotion movement has created or exacerbated a number of undesirable developments: devotion of substantial resources to an ever-expanding search for risk factors; premature exhortation of the public to adopt numerous alterations of lifestyle, with frequent reversals of advice, disregard for untoward concomitants, and exaggerated promises (explicit and implicit) concerning what such behaviors would achieve for the adopters; a public generally confused, anxious, and skeptical with regard to public-health advice; a scientific community which rushes tentative findings into print, and a mass media establishment which abets and sometimes worsens the problem; and an introspective approach to health which fosters victim blaming and stigmatization, ignores critical social, economic, and environmental issues which have major impacts on health, and further encourages an unhealthy level of concern for personal, rather than societal, well-being.

Mounting a successful attack on these problem will require us to undertake important changes in our own (professional) lifestyles. First, we should make more effective and efficient use of our resources and regain public confidence by concentrating on a few areas in which deleterious effects on health are clear and profound (e.g., smoking prevention/cessation, alcohol and drug abuse, gross obesity, hypertension). Second, we must be draconian in deciding what we know and do not know, and more conservative in our public promises of the economic and health-related benefits of our recommendations. Third, we have a duty to describe risk factors in terms of absolute as well as relative risk; to attempt to fully describe and quantify the risks of the behaviors we recommend; and to inform the public that, since there is always a "down side" to any new or altered behavior, they have to make a cost-benefit choice.

Fourth, we should persuade the media of the need to focus on research trends; to use multiple sources of information; to be critical of what they report; and to understand the limitations of scientific investigation. We must similarly educate the public concerning the notion that science deals with probabilities and not absolutes; that progress is incremental; and that one is not generally advised to undertake radical behavioral changes based on one or a few studies. Finally, while such personal choices as smoking or a highly sedentary lifestyle may indeed pose dangers to health, we must turn more attention and resources to the larger context of influences on health status and mortality. A small increase in education or economic level for an individual or population has a much greater impact on health than all of our health resources combined (Levin, 1987). As Yankauer (1984) notes, "Only the future can tell us if the current enthusiasm for prevention is truly a cultural condition or is only a screen of rhetoric that prevents us from facing more basic problems of the civilization of our time" (p. 326).

If the health-promotion movement can adopt a philosophy of well-defined and constrained focus, reasonable scientific consensus, realistic goals and claims, and willingness to acknowledge and confront the macro-context of (and influences on) health and well-being, the movement's extant contributions will be multiplied manyfold.

REFERENCES

Ahrens, E. H., Jr. (1979). Dietary fats and coronary heart disease: Unfinished business. *Lancet, 2*, 1345–1348.

Ahrens, E. H., Jr. (1985). The diet-heart question in 1985: Has it really been settled? *Lancet, 2*, 1085–1087.

Alonzo, A. A. (1985). Health as situational adoption: A social psychological perspective. *Social Science and Medicine, 21*, 1341–1344.

Avogaro, P. (1985). Apolipoproteins, the lipid hypothesis, and coronary heart disease. In R. M. Kaplan & M. H. Criqui (Eds.), *Behavioral epidemiology and disease prevention* (pp. 57–65). New York Plenum.

Barsky, A. J. (1988). The paradox of health. *New England Journal of Medicine, 318*, 414–418.

Becker, M. H. (1986). The tyranny of health promotion. *Public Health Reviews, 14*, 15–25.

Becker, M. H. (1987). The cholesterol saga: Whither health promotion? *Annals of Internal Medicine, 106*, 623–626.

Berwick, D. M. (1985). Screening in health fairs: A critical review of benefits, risk and costs. *Journal of the American Medical Association, 254*, 1492–1498.

Bloom, B. S., & Soper, M. A. (1986). Diagnostic testing for coronary artery disease in a large population. *American Journal of Preventive Medicine, 2*, 35–41.

Brown, E. R., & Margo, G. E. (1978). Health education: Can the reformers be reformed? *International Journal of Health Services, 8*, 6–9.

Carlyon, W. H. (1984). Disease prevention/health promotion—bridging the gap to wellness. *Health Values: Achieving High Level Wellness, 8*, 27–30.

Coreil, J., & Levin, J. S. (1985). A critique of the life style concept in public health education. *International Quarterly of Community Health Education, 5*, 103–114.

Coronary Drug Project Research Group. (1970). The Coronary Drug Project: Initial findings leading to modifications of its research protocol. *Journal of the American Medical Association, 214*, 1303–1313.

Eckholm, E. (1985, August 6). Health benefits of lifelong leanness are challenged by new weight table. *New York Times*, pp. 19–20.

Eisenberg, L. (1977). The perils of prevention: A cautionary note. *New England Journal of Medicine, 297*, 1230–1232.

Etzioni, A. (1972, June 3). Human beings are not very easy to change after all. *Saturday Review*, pp. 45–47.

Feinstein, A. R., & Esdaile J. M. (1987). Incidence, prevalence, and evidence: Scientific problems in epidemiologic statistics for the occurence of cancer. *American Journal of Medicine, 82*, 113–123.

Fihn, S. D. (1987). A prudent approach to control of cholesterol levels. *Journal of the American Medical Association, 258*, 2416–2418.

Frank, J. W. (1985). Occult-blood screening for colorectal carcinoma: The risks, *American Journal of Preventive Medicine, 1*, 25–32.

Freudenberg, N. (1981). Health education for social change: A strategy for public health in the U.S. *International Journal of Health Education, 24,* 138–145.

Gillick, M. R. (1984). Health promotion, jogging, and the pursuit of the moral life. *Journal of Health Politics, Policy, and Law, 9,* 369–387.

Grundy, S. M., Bilheimer, D., Blackburn, H., Brown, W. V., Kwiterovich, P. O., Mattson, F., Schonfeld, G., & Weidman, W. H. (1982). Rationale of the diet-heart statement of the American Heart Association: Report of Nutrition Committee. *Circulation, 65,* 839A–854A.

Gunning-Schepers, L. J., & Hagen J. H. (1987). Avoidable burden of illness: How much can prevention contribute to health? *Social Science and Medicine, 24,* 945–951.

Gurin, J. (1984, October). The us generation. *American Health,* pp. 40–41

Harris, L. (1987). *Inside America.* New York: Vintage Books.

Hirsch, J. (1986, May/June). Family nutrition guide. *Rx Being Well,* pp. 9–10.

Job, R. F. S. (1988). Effective and ineffective use of fear in health promotion campaigns. *American Journal of Public Health, 78,* 163–167.

Kaplan, R. M. (1985). Behavioral epidemiology, health promotion, and health services. *Medical Care, 23,* 564–583.

Keys, A. (1970). Coronary heart disease in seven countries. *Circulation, 41,* (Suppl. 1).

Koplan, J. P., Powell, K. E., Sikes, R. K., Shirley, R. W. & Campbell, C. C. (1982). An epidemiological study of the benefits and risks of running. *Journal of the American Medical Association, 248,* 3118–3121.

Koplan, J. P., Siscovick D. S., & Goldbaum G. M. (1985). The risks of exercise: A public health view of injuries and hazards. *Public Health Reports, 100,* 189–195.

Kraus, J. F., & Conroy C. (1984). Mortality and morbidity from injuries in sports and recreation. *Annual Review of Public Health, 5,* 163–192.

Kruse, D. L., & McBeath A. A. (1980). Bicycle accidents and injuries. *American Journal of Sports Medicine, 8,* 342–344.

Langone, J. (1985). The annual physical re-examined. *Discover, 6,* 50–52.

Levin, L. S. (1987). Every silver lining has a cloud: The limits of health promotion. *Social Policy, 27,* 57–60.

Lipton, J. P., & Hershaft, A. M. (1985). On the widespread acceptance of dubious medical findings. *Journal of Health and Social Behavior, 26,* 336–351.

Love, R. R. (1985). The efficacy of screening from carcinoma of the prostate by digital examination. *American Journal of Preventive Medicine, 1,* 36–46.

Medical Research Council, Coronary Drug Research Group (1975). Clofibrate and niacin in coronary heart disease. *Journal of the American Medical Association, 231,* 360–381.

Milio, N. (1981). *Promoting health through public policy.* Philadelphia: F. A. Davis Co.

National Institutes of Health Consensus Conference. (1985). Lowering blood cholesterol to prevent heart disease. *Journal of the American Medical Association, 253,* 2080–2090.

Neaton, J. D., Broste, S., Cohen, L., Fishman, E. L., Kjelsberg, M. O., & Schoenberger, J. (1981). The multiple risk factor intervention trial (MRFIT): VII. A comparison of the risk factor changes between the two study groups. *Preventive Medicine, 10,* 519–543.

O'Malley, M. S., & Fletcher, S. W. (1987). Screening for breast cancer with breast self-examination. *Journal of the American Medical Association, 257,* 2197–2203.

Palumbo, P. J. (1988). National cholesterol education program: Does the emperor have any clothes? *Mayo Clinic Proceedings, 63,* 88–90

Pooling Project Reseach Group (1978). Relationship of blood pressure, serum cholesterol, smoking habit, relative weight and ECG abnormalities to incidence of major coronary events: Final report of the pooling project. *Journal of Chronic Diseases, 31,* 201–206.

Reiser, R. (1984). A commentary on the rationale of the diet-heart statement of the American Heart Association. *American Journal of Clinical Nutrition, 40,* 654–658.

Skrabanek, P. (1988a). Benefits of mass breast screening rest on equivocal evidence. *Diagnostic Imaging, 10,* 73–77, 183.

Skrabanek, P. (1988b). Cervical cancer screening: The time for reappraisal. *Canadian Journal of Public Health, 79,* 86–89.

Slater, C., & Carlton, B. (1985). Behavior, lifestyle, and socioeconomic variables as determinants of health status: Implications for health policy development. *American Journal of Preventive Medicine, 1,* 25–33.

Stallones, R. A. (1983). Ischemic heart disease and lipids in blood and diet. *Annual Review of Nutrition, 3,* 155–185.

Stein, H. F. (1982). Neo-Darwinism and survival through fitness. *Journal of Psychohistory, 10,* 163–187.

Stunkard, A. J. (1978). Behavioral treatment of obesity: The current status. *International Journal of Obesity, 2,* 237–248.

Taylor, W. C., Pass, T. M., Shepard, D. S., & Komaroff, A. L. (1987). Cholesterol reduction and life expectancy: A model incorporating multiple risk factors. *Annals of Internal Medicine, 106,* 605–614.

Thomas, L. (1983). An epidemic of apprehension. *Discover, 4,* 78–80.

Underwood, B. A. (1986). The diet-cancer conundrum. *Public Health Reviews, 14,* 191–210.

U.S. Office of Technology Assessment. (1982). *Cancer risk: Assessing and reducing the dangers in our society.* Boulder, CO: Westview Press.

Virkkunen, M. (1985). Lipid Research Clinics coronary primary prevention trial results letter. *Journal of the American Medical Association, 253,* 635–636.

Wikler, D. (1987). Who should be blamed for being sick? *Health Education Quarterly, 14,* 11–25.

Winsten, J. A. (1985). Science and the media: The boundaries of truth. *Health Affairs, 4,* 5–23.

Yankauer, A. (1984). Prevention and the public health. *Preventive Medicine, 13,* 323–326.

15

Healthy People—Healthy Business: A Critical Review of Stress Management Programs in the Workplace*

Kenneth R. Pelletier
University of California School of Medicine

Robert W. Lutz
University of California School of Medicine

BACKGROUND

Stress is widely recognized by health professionals, public policy makers, and corporate medical planners as a significant health factor. It is estimated that 60% to 90% of visits to health-care professionals are for stress-related disorders (Cummings & VandenBos, 1981; Elite, 1986). Both basic and clinical experimental research have determined stress to be a major factor in a wide range of conditions including hypertension, cardiovascular disease, gastrointestinal disorders, tension and vascular headaches, low-back pain, and decreased immunological functioning with its implications for susceptibility to disorders ranging from colds and flus to cancer and AIDS (Pelletier, 1977; Pelletier & Herzing, 1988).

Stress has been demonstrated to also affect health-related behaviors such as cigarette smoking, alcohol use, work productivity, and absenteeism rates (McLeroy, Green, Mullen & Foshee, 1984; Seamonds, 1983). Furthermore, stress has long been implicated in the development of mental disorders and decreased life satisfaction (Pelletier, 1977, 1984, 1986). Estimates of the impact of stress disorders in financial terms to business are that a minimum of 150 billion dollars are lost annually in industrial costs due to decreased productivity, absenteeism, and disability. With national health care exceeding one

---

*This chapter is based on an article with the same title. Copyright 1988, *American Journal of Health Promotion*. Reprinted with permission. This chapter was intended to be placed as Chapter 9, but for technical reasons, appears as Chapter 15.

billion dollars daily, it is apparent why the Surgeon General's 1979 report "Healthy People" identified stress as in a major priority in the formation of a national prevention strategy through 1990.

WORKSITE STRESS MANAGEMENT PROGRAMS

There are obvious and compelling reasons for the business community to take a more active role in promoting employee health. Of the $640 billion 1989 budget for medical care, private corporations control approximately 35% to 40% of that total through the medical plans they purchase. Future trends are likely to increase that percentage to as much as 70% or higher due to the aging of the workforce, cost shifting from government to the private sector, unfunded retirement plan liabilities, extended long-term care, expanded mandatory benefits, new medical technologies, and national health insurance proposals. Medical care costs and benefits are currently about one third of the average person's salary and are continuing to rise at a rate that is two to three times greater than the overall consumer price index. Cardiovascular disease alone was estimated in 1963 to account for 12% of work days lost, and represented an economic loss of $4 billion (Felton & Cole, 1963). AT&T recently estimated their medical expenses for a single employee suffering a heart attack at about $60,000. Data collected in the 1970s (Bonica, 1980) demonstrated chronic back disorders to cost $12 billion annually in lost work days and another $4 billion in medical costs. In 1982, work accidents cost $31.4 billion and resulted in over 12,000 deaths (Jones & Dosedel, 1986). Stress has been directly linked to each of the above medical disorders, and inferentially with a host of others. Indirect costs related to stress include reduced productivity, absenteeism, job dissatisfaction, terminations, and litigation for stress disabilities (Brodsky, 1984). Reports from the Bureau of National Affairs (Bureau of National Affairs, 1984) indicate that about 50% of worker absenteeism was avoidable through appropriate attention to the physical and emotional need of employees. Given these cost incentives, and the literature demonstrating decreased health costs associated with mental health care, it is not surprising that many companies have been motivated to implement onsite stress-management programs.

Although the focus of this chapter is upon objective assessment of health and cost outcome, there are numerous worksite stress management programs that are justified by sponsoring corporations based solely on indirect benefits. In these instances, the direct benefits of health and/or cost efficacy are simply not deemed to be important. Among the reported indirect benefits are enhanced employee morale, improved corporate image, ability to attract and retain key personnel, consistency of a corporate product with the image of a healthy company, and as a perk for key executives. These are legitimate reasons in their own right, but in a decade of increasing competition, mergers, takeovers, and the ubiquitous *bottom line,* it is more likely than not that such good intentions may initiate a program but not sustain it. Overall, the trend is toward objective evaluations of both health and cost efficacy in all aspects of health promotion including workplace stress management programs.

Three questions arise consistently with regard to stress management programs in the workplace: (a) What types of interventions have been used? (b) Where are the objective studies demonstrating efficacy? and (c) Are such interventions cost-effective? Numerous national surveys have indicated that stress-management programs and their outcomes are the foremost concern of employees, employers, researchers, and science writers (Wang, et al., 1987). One survey of 164 corporate health programs indicated that stress management programs are cited four times more frequently as a priority for development than the next closest category which is the behavioral management of coronary heart disease (Pelletier, 1984). This observation was later confirmed in a survey sponsored by the U.S. Office of Disease Prevention and Health Promotion (Windom, McGinnus, & Fielding, 1987). Clearly, stress management programs and their health and cost efficacy are major areas of corporate and academic concern.

To address this complex area, it is essential to provide a context. Because it is not the intent of this chapter to review the general area of stress management interventions, it is important to point out that there is unequivocal, clinical, and experimental evidence for a wide range of interventions that can successfully alleviate stress-related disabilities. Among the multiple interventions are: meditation (Shapiro & Walsh, 1984); clinical biofeedback (Shapiro, Stoyuva, Kamrya, Barber, Miller, & Schwartz, 1979/1980); Autogenic Training (Luthe, 1965); Jacobson's Progressive Relaxation (Bernstein & Borkover, 1973); hypnosis (Kroger, 1963); the Relaxation Response (Benson, 1979); and visualization therapies (Boysenko, 1987). Despite the impressive and growing evidence of clinical efficacy, the three questions noted earlier regarding the transfer of such intervention into the workplace remain largely unanswered. One of the many reasons why this is a critical issue is that corporate health-promotion programs frequently adopt stress management programs based upon the claims of ubiquitous vendors that their prepackaged interventions have proven efficacy. Given this claim, it is essential to point out that these unfounded claims are based solely upon clinical experimental data derived outside the workplace. For this chapter, "prepackaged" programs refer to a predetermined, nonindividualized program or course in stress management most often delivered by nonhealth care professionals. There is no evidence at the present time that prepackaged stress management programs are effectively transferred into the work environment.

Finally, there are two other contextual issues to be considered. Stake are very high in developing lifestyle interventions to prevent chronic disease and disability. The Carter Center report "Closing the Gap" (The Carter Center of Emory University, 1985) indicated that given our present medical knowledge and capabilities, approximately two thirds of all deaths in this country are premature, and further, that about two thirds of all years of life lost before age 65 are preventable. These are astounding statements especially given the fact that prevention-related research is allocated less than 5% of overall health funding in the United States. Translated into economic terms, the projected medical budget for 1989 is estimated to exceed half a trillion dollars ($610 billion), or 11.4% of the Gross National Product.

It seems evident that far greater expenditures could be productively allocated to developing clinical experimental protocols for behavioral medicine interventions to alleviate premature morbidity and mortality. Furthermore, determining the areas for the greatest efficacy is certainly no mystery. Writing in the *Journal of the American Medical Association,* Assistant Secretary of the U.S. Department of Health and Human Services, William H. Foege, has clearly stated:

> The scientific basis for the influence of life-style choices on health continues to grow; lifestyles are changing and it is probably that these changes are already reducing the toll of diseases. In the coming decades, the most important determinants of health and longevity will be the personal choices made by each individual. This is both frightening for those who wish to avoid such responsibility and exciting for those who desire some control over their own destiny (Foege, 1985).

Finally, the overall field of health-promotion programs in the workplace, of which stress management would be only one aspect, has not been adequately evaluated by rigorous design and appropriate data analysis (Elias & Murphy, 1986). Given the lack of such analytic work, it is therefore not surprising that stress management per se has not been adequately evaluated. To date, there are only nine studies of sufficient precision that evaluate comprehensive health-promotion programs and present evidence of efficacy. These few studies are in obvious sharp contrast to overly zealous claims by some vendors of health-promotion programs and stress management interventions, in particular.

Among the rigorous studies of overall health-promotions programs to date are those from AT&T (Spilman, Goetz, Schultz, Bellingham, & Johnson, 1986), Prudential Insurance (Bowne, Russell, Morgan, Optenberg, & Clarke, 1984), Canada Life and North American Life (Shephard, Corey, Ruezland, & Cox, 1982), Tenneco (Baun, Bernacki, & Tsai, 1986), Blue Cross and Blue Shield of Indiana (Gibbs, Mulvaney Henes, & Reed, 1985), Blue Cross of California (Lorig, Kraines, Brown & Richardson, 1985), Control Data Corporation (Jose & Anderson, 1986), and two studies from Johnson & Johnson demonstrating the health efficacy of a systems approach (Blair, Collingwood,

Reynolds, et al., 1984) and overall positive impact of the program in comparative worksites (Bly, Jones, & Richardson, 1986). Eight of the nine studies demonstrated clear efficacy in improving health, with one exercise program (Baun, et al., 1986) evidencing equivocal results. Cost reduction is evident in three (Blair, Collingwood, Reynolds, et al., 1984; Bly, Jones, & Richardson, 1986; Gibbs, Mulvaney, Henes, & Reed, 1985) out of the nine although the most recent Johnson & Johnson study has been challenged on both methodological grounds and on cost efficacy because program costs were not included in the final analysis.

Most significantly for this chapter is that only the 1984 Johnson & Johnson study reported a positive objective outcome of improved stress management by employees. None of the nine studies isolated and assessed the specific impact of the stress management component of the respective programs. In any consideration of the nature, health outcome, and cost efficacy of workplace stress-management programs, it is essential to bear in mind this context as well as these limitations in the data.

Although the focus of this review is upon workplace stress management programs, it is necessary to briefly cite the nonworkplace research because it is more extensive and has a direct bearing on questions of the relative efficacy of worksite programs.

Clinical Stress Management

Clearly, the magnitude of stress-related disorders in terms of its cost to society, as well as human suffering is enormous. The tragedy, or promise, depending on your perspective, is that many stress-related health problems are preventable. Data supporting the reversibility of rising stress-related health care costs comes largely from studies by Health maintenance Organizations (HMO's) and Third Party Carriers (TPC's). These studies primarily addressed the impact of brief outpatient psychotherapy in medical settings. Although brief psychotherapy may seem a large step away from the typical stress management program offered in worksites, it is exactly this difference in formatting (and resulted outcomes) that warrants close examination. This is particularly true because these objectively efficacious studies are often cited to support implementing prepackaged worksite stress-management programs.

An early review article by Jones and Vischi (1979) analyzed 13 studies that focused on the relationship between mental health services and medical care utilization. Reductions in medical utilization of 5% to 85% were reported in 12 of the 13 studies. One study that reported an increase rather than a decrease in medical use, stemmed from data collected in a new neighborhood health center where medical needs were previously underserved and could thus be expected to rise greatly with new services offered. In a later paper, Mumford, Schlesinger, and Glass (Mumford, Schlesinger, & Glass, 1982) reviewed 13 studies evaluating the effects of psychological intervention on recovery from surgery and heart-coronary occurrences. Virtually all of the studies used very brief, psychotherapy oriented interventions, yet were able to reduce hospitalization time by an average of 2 days for experimental patients versus control groups.

An earlier study by Jameson, Shuman, and Young (1978) examined the effects of fee-for-service outpatient psychiatric care on overall costs of third-party coverage. One hundred thirty-six Blue Cross claims records of persons using mental health services were examined over a 48-month period. Of these, 83% had 15 or less psychological contact services. Analysis of their pre-versus postpsychotherapy claims showed that their overall cost in claims for medical/ surgical utilization dropped by more than 50% following psychotherapy. In a similar but more recent study, Schlesinger, Mumford, Glass, Patrick, and Scharfstein (1983) examined medical care costs of patients covered by Blue Cross/ Blue Shield who were diagnosed which chronic diseases, and compared those who subsequently received mental health services with those who did not. Persons receiving psychological services had medical charges significantly lower than the comparison group. They conclude that mental health services combined with medical services can both improve the quality of care and lower provider costs.

Health Maintenance Organizations theoretically share with the business world a direct fiscal incentive to keep members healthy in order to contain overall medical costs. Despite this apparent incentive, HMO's have tended to function nearly identically to fee-for-service providers in terms of having a disease-treatment approach rather than a ''health maintenance'' or disease prevention and health-promotion orientation. Factors such as self-selection by younger and healthier patients and possible underutilization of diagnostic and invasive procedures may be responsible for perceived HMO cost-effectiveness. With these caveats in mind, it is useful to examine the recent experience of the Kaiser-Permanente Health Plan which is the forerunner of the modern HMO. Reporting on the 20-year experience of Kaiser in assessing psychotherapy and medical utilization, Cummings and VandenBos (1981) note that 60% to 90% of stress-related medical costs are more than offset by appropriate psychological interventions. Based on this major study, they state:

> The findings indicated that: (a) persons in emotional distress were significantly higher users of both inpatient (hospitalization) and outpatient medical facilities as compared to the health plan average; (b) there were significant declines in medical utilization by those emotionally distressed individuals who received psychotherapy, as compared to the control group; (c) these declines remained constant during the five years following the termination of psychotherapy; (d) the most significant declines occurred in the second year after the initial interview, and those patients receiving one session only or brief psychotherapy (two to eight sessions) did not require additional psychotherapy to maintain the lower level of medical utilization for five years . . . . (pp. 63–74)

These conclusions are further bolstered by more recent data from a pilot project in Hawaii demonstrating a 37% decrease in medical use following brief psychotherapy (Cummings, 1985). In sum, data for the efficacy of brief psychotherapy to address stress-related symptoms and medical care costs is quite well established. However, the questions of whether or not such approaches and outcomes are transferable to the corporate world are unanswered at present.

EFFICACY OF WORKSITE STRESS PROGRAMS

To date, the best *methodological* critique of workplace stress management programs is that of McLeroy, Green, Mullen and Foshee (McLeroy et al., 1984) who critiqued the methodological adequacy of 19 worksite stress management programs. Among the studies reviewed were those located in public and private institutions serving both blue collar and white collar workers. Sixteen of the 19 studies used volunteer subjects as opposed to selecting subjects on the basis of diagnosed stress symptoms or "at risk" status as in the psychotherapy/medical care utilization studies previously described. The majority of worksite studies used individual sessions of training in stress management, although very few individualized such training.

Standardized interventions were the rule. Relaxation and/or meditation training was provided in 80% of the programs and half of these studies used this as the sole stress management method technique. Other studies combined some form of relaxation training with either biofeedback or cognitive restructuring techniques. Virtually all programs provided participants with information about the nature and importance of stress along with the stress management training. Stress *reduction* methods such as assertiveness training or increasing employees' actual work options were used in only 3 of the 19 studies. Approximately two thirds of the studies provided stress-management training on a once-a-week basis for 4 to 8 weeks. Only 8 of the 19 studies reported any follow-up data whatsoever. These generally noted some deterioration in treatment effects that appeared partly related to lack of carry-over and/or long-term compliance with the stress management exercises.

Due to the methodological focus taken in the review by McLeroy and colleagues (McLeroy et al., 1984), the many instances of negative findings on dependent measures for stress reduction were not well defined. Although most worksite studies that were reviewed reported some positive effects, overall they were inconsistent and not very compelling. Only 4 of the 19 studies directly measured cost-affected dependent variables. Manuso (1983) reported two studies measuring changes in use of health services. Seamonds (1982) and Jackson (1983) reviewed company records to assess absenteeism. Of these studies, Manuso and Seamonds were the only 3 studies of the 19 reviewed by McLeroy to select subjects on the basis of stress symptoms or "at risk" status.

Of the 15 remaining studies, 10 indicated at least on objectively measured physiological dependent variable. Five studies measured blood pressure, 4 measured EMG and/or peripheral temperature circulatory changes, and 1 included urine lab values. Non related the physiological outcome data to subsequent medical care use. Thirteen of the 19 studies included various paper-and-pencil tests to measure variables such as job satisfaction, coping effectiveness, stress level, anxiety, depression, and somatic complaints. Ten of the 19 studies employed experimental designs with random assignment of subjects to experimental and control conditions.

Clearly, the most consistently reported findings appear to be in reducing subjective feelings of emotional distress such as anxiety, although even in that case, the results are certainly not highly consistent. Reductions in somatic complaints (other than blood pressure) are inconsistent across studies, and those studies obtaining reductions tend to observe significant regression in follow-up data. This is in sharp contrast to the HMO studies cited earlier that show increasing treatment gains over time. Job satisfaction indices show most inconsistency with one study actually reporting increased job dissatisfaction after stress management training (Murphy, 1982). This latter finding is almost certainly linked to the fact that virtually all programs educate participants as to the sources and effects of stress, yet almost none attempt to actually reduce work stresses. Increasing worker sensitivity to job stress and then not offering any work modification options is clearly not an advisable strategy to increase job satisfaction for most employees. Such a strategy results in the unfortunate paradox of "healthy people in unhealthy places" (Pelletier, 1984).

DIFFERENCES BETWEEN WORKSITE AND CLINICAL PROGRAMS

HMO and TPC studies demonstrating cost effectiveness for stress reduction rely on three factors: (a) effective identification of symptomatic persons; (b) Thorough evaluation of identified persons by health professionals who then make recommendations and/or referrals; and (c) Reliance on appropriate, brief, symptom oriented interventions when indicated, most often in the form of brief psychotherapy. Brief psychotherapy itself share many commonalities with *one-on-one*, professionally conducted, stress management therapies, and may be synonymous in any instances. One-on-one stress management may more commonly involve the learning of at least one specific relaxation and/or cognitive stress management skill. On the other hand, brief psychotherapy may more clearly devote the importance of the therapeutic relationship and attention to broader individual differences and needs. However, both one-on-one stress management therapy and brief psychotherapy involve a focused, time limited and stress alleviating orientation to defined problems under the

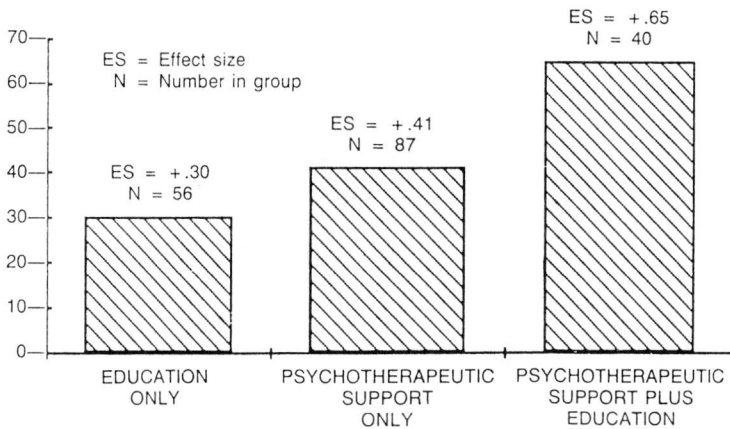

Figure 15.1.   Average outcome effect size in improving recovery to surgery and heart attack.

care of an appropriately trained, health care professional. It would seem that the combination of brief psychotherapy and stress management skills would provide an optimal approach to stress related disorders. Relevant to this assertion, Mumford, Shlesinger and Glass (1982) noted that although both educational and psychotherapeutic interventions were effective by themselves in improving the recovery of patients following surgery and heart attacks, they were still more effective when used in combination. Figure 15.1 shows this summary data. In sharp contrast to brief psychotherapy, most worksite stress management programs offer employee volunteers a predetermined, non-individualized stress management package. There is a huge discrepancy in the questionable and limited outcome data generated from most pre-packaged worksite stress management programs with the previously described brief psychotherapy data generated by HMO's and TPC's. Further, this discrepancy appears to reflect a systematic difference in the manner in which stress management interventions are conducted.

MISLEADING ASSUMPTIONS

Common use of relatively simplistic stress management training modules in worksites is most likely, a reflection of their marketability due to their apparent low cost and reported efficacy. Unfortunately, this approach seems to reflect a misunderstanding if not misinterpretation of what works in lowering stress and stress-related medical costs. Relaxation and biofeedback training used in clinical practice have been shown to decrease stress/tension-related somatic symptoms, and also to significantly reduce excessive or inappropriate

197

medical utilization (Manuso, 1983). Clearly, these interventions do have certain "generic" components. Therefore, it is not unreasonable to assume that these techniques could be transplanted from their clinical context, developed into a standardized but flexible program format, and applied as a generic intervention for stress disorders. However, this appears to be an erroneous assumption. Although the development of such programs may be theoretically possible and a plethora of vendors make claims based on this misleading assumption, it is evident that empirically effective prepackaged stress management programs do *not* exist at the present time, so "caveat emptor."

Manuso's (1983) study of a workplace stress management program provides a good illustration. It is one of the most frequently cited studies in the literature, and one of the few worksite stress-intervention studies to date to report a dollar-and-cents cost-benefit index. It is also commonly over simplified in reviews to suggest direct support for prepackaged programs. Murphy (1982) describes Manuso's study in support of the efficacy of biofeedback:

> Several recent studies indicate that biofeedback training can be effective in worksite stress management training programs. Biofeedback training has been employed along with breathing exercises and muscle relaxation in an emotional health program at the Equitable Life Assurance Society to help 30 employees cope with chronic anxiety and recurring headaches. Training sessions were held three times per week over a five week period. Participants were able to reduce muscle tension levels by 50% after training and variability of response by 600%. At follow-up three months later, participants reported fewer visits to the health center, decreases in stress symptoms, increases in work satisfaction, and decreases in symptom interference with work. Cost/benefits analysis indicated an impressive 1:5.5 ration for the training program.

Simplifying the Manuso (1983) study still more, the review by McLeroy et al. (1984) describes Manuso's intervention as, "a five week program of biofeedback and muscle relaxation."

In actuality, Manuso's (1983) study is quite unlike prepackaged worksite stress programs. Because it is so different in format and outcome, it warrants further consideration. In the Equitable program, Manuso (1983) reports a four-stage intervention. First, employees with a long history of a stress-related disorders were identified and referred for evaluation. He notes that they were then screened medically, neurologically, and psychologically to assure their suitability. Employees also had to show their motivation and complete home recordkeeping. Second, baseline measurements were obtained over a 2-week period involving biofeedback readings and daily self-monitoring logs. In the treatment phase, individuals were scheduled for two to three training sessions per week over 5 weeks and were helped to wean themselves from reliance on instrumental-aided autoregulation. They also were given a number of home exercises to increase treatment generalizability. Finally, followup provided once or twice a year included baseline reassessments and advice to maximize treatment affects. To describe this clinical screening and intervention trial simply as biofeedback greatly obscures what was done, and lends itself to the misperception that simple interventions can produce dramatic and cost-effective results. Manuso's 1983 study is one of only three studies reviewed by McLeroy et al. (1984) that reported positive, cost-relevant outcome measures. Most significantly, both of the other studies, a companion study by Manuso (1983) on Type A versus Type B employees, and a study showing decreased absenteeism following individual counseling, also used thorough evaluation and identification of symptomatic or high-risk subjects.

A recent and most important study lends further support to the greater efficacy of individualized program with targeted participants over more common prepackaged stress training in the workplace. Reporting in the *American Journal of Health Promotion*, Steffy and his colleagues (1986) reported on organizational stress management to reduce employee accidents and consequent costs. Their approach was applied in three settings including a medium-sized hospital, a trucking company, and a small hospital. The authors identified five general components to their overall interventions: (a) Increasing the awareness of senior management to stress in their organization and its relationship to insurance losses; (b) Feedback to management on their assessment regarding the sources and levels of stress as a function of location, department, job shift, and demographics; (c) Identification of high-risk groups/units, along with personal consultations to correct underlying problems: (d) Establishing an Employee Assistant Program to provide ongoing psychological services to employees and their families for work and nonwork problems; and (e) Various health programs typically including stress management training and educational inducements toward exercise, healthy lifestyle management, and avoiding stress-related back injuries.

All three intervention sites in the Steffy et al. (1986) study showed significantly reduced accident claims and costs. Combined hospital accident-related losses went from an average of $24,199 per month in the 2 years preintervention to $2,577 per month in the 11 months postintervention. One hospital reduced the number of claims from 3.1 per month to 0.6 and the average monthly cost from $7,329 to $324. Cost of the program relative to the retrieved losses were not reported. In general, when such ratios are analyzed, the costs for stress management interventions are more than offset by the savings in disability costs and lost productivity.

CONCLUSIONS

Stress management seminars and prepackaged training that generally involve some form of relaxation skill carry a relatively low fee and are logistically easy to implement within most work settings. However, such simple interventions should not be expected to significantly alter stress-related costs of medical care or have a positive impact on overall productivity. It may be worthwhile to consider the few common elements that the most efficacious programs share before considering a fundamentally different and more effective model.

Based on the methodological critique of McLeroy et al. (1984), a position paper prepared for the Washington Business Group on Health (Jaffee, Scott, & Orioli, 1986), and two previous critical reviews (Pelletier, 1986; Pelletier, Lutz, & Klehr, 1987), several elements appear in common and may serve as the basis for the development and evaluation of future, prepackaged programs that may actually be of demonstrable benefit. Among these aspects are (a) Complete programs take place once a week for 4 to 8 weeks; (b) Sessions are usually 45 min. in length; (c) An ideal number of participants range from 12 to 15; (d) Peer-led programs appear to be equally effective or ineffective as professionally led interventions; (e) All programs taught participants one or more stress management techniques such as abdominal breathing and/or assertiveness training; (f) Because none of the studies indicated continued practice at 3 to 6 months, programs should include homework, generalization skills, and periodic followup; (g) Evaluation to determine a program's efficacy; and (h) Ideally, the workplace environment needs to correct its inherently stressful nature to avoid the frequent "blame the victim" orientation. Although those characteristics are approximations in the absence of hard data, they do provide an initial set of criteria by which more effective programs may be developed and evaluated.

Below are outlined the key components common to successful (health and cost effective) stress management programs applied in worksites and HMO's alike:

- Identification of symptomatic and high risk individuals;
- Thorough evaluation by qualified health professionals to assess both the medical and psychological status and needs of high-risk individuals;
- Appropriate referral or treatment of motivated individuals. Applied treatments not to be limited to work stressors or a strict definition of stress management;
- Treatments are symptoms oriented and provided by professionals skilled with brief intervention protocols;
- Appropriate followup is provided to assure treatment maintenance and generalization.

Evaluations are conducted to determine health and/or cost efficacy. Based on the above analysis, the Steffy et al. study (1986) provides a *potential* model for effective stress reduction/health promotion in the workplace. Their program includes many of the key program components cited in this literature review, but orients them more specifically toward the workplace. Thus, identification of high-risk organizational units are emphasized in addition to individuals. Problems are also addressed through systems-level organizational changes, as well as individually oriented treatments and referrals. Unfortunately, these authors have not yet reported the effects of their program on reducing medical care utilization, absenteeism, or toward increasing productivity. Further research needs to address these questions, as well as whether such systems-level intervention additions provide significantly more benefits over individually oriented programs so as to remain cost effective. Finally, the magnitude of the problem of potentially preventing two thirds of all morbidity and mortality before age 65 and the reallocation of in excess of $610 billion is simply overwhelming. Consensus indicates that lifestyle interventions such as effective stress management programs are the cornerstone of such efforts. However, it is critical to underscore that the resolution of such a colossal undertaking is dependent upon the effective coordination of the three basic elements of a true health care system: (a) Quality, cost-effective medical care; (b) Cost containment and utilization review procedures to guard against over or underutilization; and (c) Quality, cost-effective disease prevention, health promotion, and behavioral medicine programs. At the present time, there are a few promising instances of addressing these three elements in HMO and corporate settings. One example is an innovative medical plan termed "Customcare" at Southwestern Bell Corporation in St. Louis that contains all three elements and is the subject of a coordinated research effort by Southwestern Bell, Prudential Insurance, and Johnson & Johnson Health Management Incorporated (Fielding & Pelletier, 1987). This evaluation is scheduled for completion in early 1991.

There is an ongoing program initiated in 1984 at the University of California School of Medicine in San Francisco involving faculty from the Department of Medicine and representatives from 15 major corporations including AT&T, Apple Computer, Bank of America, Levi-Strauss, Lockheed, Hewlett Packard, Southwestern Bell Corporation, Wells Fargo Bank, and the Office of Prevention of the State of California (Pelletier, Klehr, & McPhee, 1987). In essence, the project is designed to develop and evaluate an array of innovative medical, cost containment, and health-promotion programs on an objective basis. Additionally, a program has been developed for the State of California Department of Mental Health (Pelletier, Lutz, & Klehr, 1988) as a self-care stress management program under the title *That's Life.* The main objective is to deliver and evaluate a stand alone self-care stress management program which is available to residents of the state of California and eventually to residents of other states as well. This program consists of print materials and audiotapes aimed for intermediate to low-literacy populations. It may be possible to modify this program for use within the workplace but this would re-quire rigorous development and evaluation. Finally, because there is no evidence that a freestanding, self-care stress program is efficacious per se, the program is clearly tied to eliciting self referrals to state and private mental health care providers. This is in recognition that stress management is a com-plex problem requiring a system resolution, not a simplistic panacea.

In these instances, stress management programs continue to be the number one priority cited by both employers and employees. Unfortunately, the com-monly used prepackaged stress management programs do not appear to be ei-ther health or cost effective at the present time based on objective assessments. Obviously, one cannot show positive health care outcomes as eas-ily if one's target audience is neither symptomatic nor at high risk, because they already have low health care costs. Here, as noted earlier, employers and employees may have reasons for desiring stress management other than health cost incentives.

At the present time, the best health and cost efficacy data indicate that stress management is most effective with an identified high risk or symptom-atic employee population utilizing traditional, problem-focused, brief psycho-therapy with an emphasis upon developing stress management skills. Whether group or individual approaches are employed, it is also essential to modify hazardous working conditions to avoid the paradox of "healthy people in un-healthy places (Pelletier, 1985)." Finally, the one unequivocal conclusion of this critical review is that program evaluation is as essential as program devel-opment. Through the collaborative research efforts of corporations and univer-sities, the pressing but unresolved issues of today should yield an abundance of insights tomorrow.

REFERENCES

Baun, W., Bernacki, E., & Tsai, S. (1986). A preliminary investigation: Effect of a corporate fitness program on absenteeism and health care costs. *Journal of Occupational Medicine, 28*, 18–22.

Benson, H. (1979). *The mind/body effect*. New York: Simon and Shuster.

Bernstein, D., & Borkover, T. (1973). *Progressive Relaxation Training*. Chicago: Research Press.

Blair, S., Collingwood, T., Reynolds, R., et al. (1984). Health promotion for education: Impact on health behaviors, satisfaction, and general wellbeing. *American Journal of Public Health, 74*, 147–149.

Bly, J., Jones, R., & Richardson, J. (1986). Impact of worksite health promotion on health care costs and utilizations: Evaluation of Johnson & Johnson's Live for Life program. *Journal of the American Medical Association, 256*(23), 3235–3240.

Bonica, J. (1980). Pain research therapy: Past and current status and future needs. In L. Ng, & J. Bonica (Eds.), *Pain, discomfort and humanitarian care*. Elsvier.

Bowne, D., Russell, M., Morgan, J., Optenberg, S., & Clarke, A. (1984). Reduced disability and health care costs in an industrial fitness program. *Journal of Occupational Medicine, 26*, 809–816.

Boysenko, J. (1987). *Minding the body, mending the mind.* Reading, MA: Addison-Wesley.

Brodsky, C. (1984). Long-term work stress. *Psychosomatics, 25*(5), 361–368.

Bureau of National Affairs (1984). *Personnel policies forum survey, 132,* 3–11.

The Carter Center of Emory University. (1985). Closing the gap: National health policy consultant. Atlanta: USGPO.

Cummings, N. (1985, June). *Saving health care dollars through psychological service? Preliminary data from pilot project.* Hawaii: Blue Cross of Hawaii.

Cummings, N., & VandenBos, G. (1981). The twenty year Kaiser-Permanente experience with psychotherapy and medical utilization. *Health Policy Quarterly, 1*(2).

Elias, W., & Murphy, R. (1986). The case for health promotion programs containing health care costs: A review of the literature. *The American Journal of Occupational Therapy, 40*(11), 759–763.

Elite, A. (1986, July). *Stress management program: RFP background paper.* Internal paper. San Francisco: California Department of Mental Health.

Felton, J., & Cole, R. (1963). The high cost of heart disase. *Circulation, 27,* 957–962.

Fielding, J., & Pelletier, K. (1987). Personal communication.

Foege, W. (1985, October). Public health and preventive medicine. *Journal of the American Medical Association, 254*(16), 2330–2332.

Gibbs, J., Mulvaney, D., Henes, C., & Reed, R. (1985). Worksite health promotion. Five-year trend in employee health care costs. *Journal of Occupational Medicine, 27,* 826–830.

Jackson, S. (1983). Participation in decision making as a strategy for reducing job-related strain. *Journal of Applied Psychology, 68,*(1), 3–19.

Jaffe, D., Scott, C., & Orioli, E. (1986). Stress management in the workplace. *Washington Business Group on Health. Worksite Wellness Series.* WBGH: Washington, DC.

Jameson, J., Shuman, L., & Young, W. (1978). The effects of outpatient psychiatric utilization on the cost of providing third-party coverage: A study sponsored by Blue Cross of Western Pennsylvania (Research Series 18). *Medical Care, 16,* 383–399.

Jones, J., & Dosedel, J. (1986). The impact of corporate stress management in insurance losses. *Legal Insight. 1*(4), 24–27

Jones, K., & Vischi, T. (1979). Impact of alcohol, drug abuse and mental health treatment on medical care utilization: A review of the research literature. *Medical Care* (supplement), *17,* ii–82.

Jose, W., & Anderson, D. (1986). Control data: The stay well program. *Corporate Commentary, 2,* 1–13.

Kroger, W. (1963). *Clinical and experimental hypnosis.* J. B. Lippincott.

Lorig, K., Kraines, R., Brown, B., Jr., & Richardson, N. (1985). A workplace health education program that reduces outpatient visits. *Medical Care, 23,* 1044–1054.

Luthe, W. (1965). *Autogenic training.,* Grune and Stratton.

Manuso, J. (1983). The Equitable Life Assurance Society program. *Preventive Medicine, 12,* 658–662.

McLeroy, K., Green, L., Mullen, K., & Foshee, V. (1984). Assessing the effects of health promotion in worksites: A review of the stress program evaluations. *Health Education Quarterly, 11*(4), 379–401.

Mumford, E., Schlesinger, H., & Glass, G. (1982). The effects of psychological intervention on recovery from surgery and heart attacks: An analysis of the literature. *American Journal of Public Health, 72*(2), 141–151.

Murphy, L. (1982). Worksite stress management programs. *Employee Assistance Programs Digest,2,* 22–25.

Murphy, L. (1983). A comparison of relaxation methods for reducing stress in nursing personnel. *Human Factors, 25*(4), 431–440.

Pelletier, K. (1977). *Mind as healer, mind as slayer: A holisitc approach to preventing stress disorders.* New York: Delecote and Delta/Seymour Lawrence.

Pelletier, K. (1984). *Healthy people in unhealthy places: stress and fitness at work.* New York: Delacorte and Delta/Seymour Lawrence.

Pelletier, K. (1985, September). White-collar health: The hidden hazards of themodern office. *The New York Times.*

Pelletier, K. (1986). Healthy people in healthy places: Health promotion programs in workplace. In M. Cataldo & T. Coates (Eds.), *Health and Industry: A behavioral medicine perspective.* Wiley.

Pelletier, K., Klehr, N., & McPhee, S. (1987). Town and gown collaboration: Development of workplace health promotion programs in process.

Pelletier, K., Lutz, R., & Klehr, N. (1987). That's life!: A self help stress management program. State of California Department of Mental Health.

Pelletier, K., Lutz, R., & Klehr, N. (1988). That's life!: A self help stress management program. State of California Department of Mental Health, San Francisco.

Pelletier, K., & Herzing, D. (1988). Psychoneuroimmunology: Toward a mindbody model—a critical review. *Advances: Journal of the Institute for the Advancement of Health, 5*(1), 27–56.

Seamonds, B. (1982). Stress factors and their effect on absenteeism in a corporate employee group. *Journal of Occupational Medicine, 24*(5), 393–397.

Seamonds, B. (1983, November). Extension of research into stress factors and their effect on illness absenteeism. *Journal of Occupational Medicine, 25*(11), 821–822.

Schlesinger, H., Mumford, E., Glass, G., Patrick, C., Scharfstein, S. (1983). Mental health treatment and medical care utilization in a fee-for-service system. *American Journal of Public Health, 73*(4), 422–429.

Shapiro, D., & Walsh, R. (Eds.). (1984). *Medication: Classic and contemporary perspectives.* Aldine.

Shapiro, D., Stoyuva, J., Kamrya, J., Barber, T., Miller, N., & Schwartz, G. (Eds.). (1979–1980). *Biofeedback and behavioral medicine.* Aldine.

Shephard, R., Corey, P., Ruezland, P., & Cox, M. (1982). The influence of an employee fitness program and lifestyle modification program upon medical care costs. *Canadian Journal of Public Health, 73,* 259–263.

Spilman, M., Goetz, A., Schultz, J., Bellingham, R., & Johnson, D. (1986). Effects of a corporate health promotion program. *Journal of Occupational Medicine, 28,* 285–289.

Steffy, B., Jones, J., Murphy, L., & Kunz, L. (1986, Fall). A demonstration of the impact of stress abatement programs on reducing employees' accidents and their costs. *American Journal of Health Promotion, 1*(2), 25–32.

Wang, P., et al. (1987, October). A cure for stress? *Newsweek,* 64–65.

Windon, R., McGinnus, J., & Fielding, J. (1987). Examining worksite health promotion programs. *Business and Health, 4*(9), 26–37.

Author Index

Numbers in *italics* denote pages with complete bibliographic information.

Felton, J., 190, *203*
Fernstrom, J. D., 94, *96*
Fielding, J., 191, 201, *203, 204*
Fielding, J. E., 14, 15, 16, 24, 25, 27, 56, 72, 79, 80, *96,* 125, 129, *131, 132,* 159, *169,* 170, 172, 173, *176*
Fihn, S. D., 179, *186*
Fisher, E. B., 90, *96*
Fisher, E. D., Jr., 110, *113*
Fishman, E. L., 180, *187*
Flay, B., 90, *96*
Fleischer, B. J., 90, *96*
Fletcher, S. W., 179, 182, *187*
Flora, J. A., 129, *131*
Foege, W., 192, *203*
Fogle, R. K., 127, *131*
Follick, M. J., 109, 110, *112, 113*
Foreyt, J. P., 99, 100, 104, 110, 111, *112, 113*
Fortmann, S. P., 129, *131,* 155, *158*
Foshee, V., 189, 195, 198, 199, 200, *203*
Fowler, J. L., 110, *113*
Fowles, D. G., 148, *153*
Fox, E., 94, *95*
Fox, S. M., 127, *131*
Frank, J. W., 182, *186*
Frazio, A. F., *27*
Freudenberg, N., 183, *187*
Friedman, M., 2, *10*
Fries, J. F., 2, *9*

Garabrandt, D. H., 118, *131*
Garfinkel, L., 78, *96*
Garner, D. T., *113*
Garrow, J. S., 124, *133*
Gavin, J. R., 121, 122, *132*
Gee, M., *113*
Gentry, W. D., 7, *9*
Gerber, W. M., 100, 103, *113*
Gibbons, L. W., 120, *131*
Gibbs, J. O., 128, *131,* 173, *176,* 192, 193, *203*
Gillick, M. R., 184, *187*
Glanz, K., 151, *153*
Glasgow, R. E., 89–90, *96, 97,* 160, *169*
Glass, G., 193–194, 197, *203, 204*
Godding, P. R., 89, 90, *96*
Goetz, A., 19, *28,* 34, *36,* 192, *204*
Goetzel, R. Z., 173, *176*
Goldbaum, G. M., 182, *187*
Goldbeck, W. B., 148, *153*
Golding, L., 124, *133*
Goodrick, G. K., 100, *113*
Goodyear, N. M., 120, *131*

Gordon, J. R., 109, *114,* 164, *169*
Gormally, J., 105, *113*
Gossard, D., 117, *131*
Gotestam, K. G., 102, 107, *113*
Gottlieb, N. H., 14, *27*
Gotto, A. M., 100, *113*
Graham, L. E., 104, 107, *113*
Green, L., 189, 195, 198, 199, 200, *203*
Grove, D. A., 90, *96*
Gruder, C. L., 90, *96*
Grunberg, N. E., 91, 92, 93–94, *96, 97*
Grundy, S. M., 179, *187*
Guire, K., 102, 104, 107, *115*
Gunning-Schepers, L. J., 180, *187*
Gurin, J., 184, *187*

Hagan, R. D., 120, 127, *131*
Hagberg, J. M., 120, 121, 122, *132*
Hagen, J. H., 180, *187*
Haight, S. A., 54, 60, 72
Hall, R. G., 102, *113*
Hall, S. M., 102, 108, *113*
Hallbauer, E. S., 101, 105, *113*
Hanson, R. W., 102, *113*
Harris, L., 184, *187*
Harris, M. B., 101, 103, 105, *113*
Hartwell, T. D., 20, *28*
Haskell, W. L., *27,* 118, 120, 123, 124, 125, 127, 129, *131*
Hattner, R. B., 122, *131*
Hayes, S. C., 108, *112,* 123
Haynes, S. G., *27*
Haynes, W. C., *132*
Heath, G. W., 121, 122, *132*
Heckerman, C. L., 108, *112*
Heckler, L. M., 90, *96*
Heinzelmann, F., 127, *131*
Heller, R. F., 90, *97*
Henderson, B. E., 78, *98*
Henderson, J. B., 90, *97*
Henes, C., 128, *131,* 173, *176,* 192, 193, *203*
Hershaft, A. M., 181, *187*
Herzing, D., 189, *204*
Herzlinger, R. E., 29, 32, *36*
Hinderlith, J. M., 121, 122, *132*
Hirsch, A. A., 108, *114*
Hirsch, J., 180, 181, *187*
Ho, J. H. C., 78, 79, *97*
Ho, P., *27*
Hollenbeck, B. B., *177*
Holloszy, J. O., 121, 122, *132*
Houston-Miller, N., 117, *131*

Subject Index